Implementing an Information Strategy in Manufacture

IFS

IMPLEMENTING AN INFORMATION STRATEGY IN MANUFACTURE

A PRACTICAL APPROACH

Jack Hollingum

IFS (Publications) Ltd, UK

Springer-Verlag
Berlin · Heidelberg · New York
London · Paris · Tokyo

Jack Hollingum
IFS (Publications) Ltd
35-39 High Street
Kempston
Bedford MK42 7BT
England

British Library Cataloguing in Publication Data

Hollingum, Jack
Implementing an information strategy in manufacture
1. Management information systems
I. Title
658.5′14 T58.64

ISBN 0-948507-47-0 IFS (Publications) Ltd
ISBN 3-540-17613-6 Springer-Verlag Berlin
ISBN 0-387-17613-6 Springer-Verlag New York

Phototypeset by Parchment (Oxford) Ltd
Printed and bound by Short Run Press Ltd, Exeter

Foreword

This book grew out of the belief that, although the potential of CIM is widely recognised, there is little understanding and great nervousness concerning the practical matter of its implementation. Are you confident that your company will find the right answers to these questions:

- In which areas can CIM most significantly influence the competitive position of your company?
- What information is it essential to include within a CIM plan?
- How can each phase of your CIM plan be cost justified, and how should it be audited for success?
- How should you set priorities for implementing the various phases of a CIM plan?
- What is the significance of networking to CIM?

Every company investing in CIM faces these questions but too frequently they are left unanswered. Our experience is that the implementation of CIM is primarily a management challenge not a technical one. The greatest challenge is how to make the organisational changes needed to obtain the benefits from CIM.

This book is a valuable guide to anyone planning to invest in CIM.

Patrick McHugh
Director, Advanced Manufacturing Technology
Coopers & Lybrand Associates
February 1987

CONTENTS

Preface

In December 1986 an event took place in Birmingham, UK, which involved communications between computer systems supporting manufacturing functions such as design and production scheduling, and with controllers of production equipment such as machine tools, robots and automated guided vehicles. This 'CIMAP' event, in which 60 companies – many of them direct competitors – collaborated under the sponsorship of the UK Government's Department of Trade and Industry, demonstrated powerfully that technology is now available to achieve what in the past has been little more than a dream of the advocates of computer integrated manufacturing: to bring together separate 'islands of information' such as computer aided design, process planning, material requirements planning, production scheduling, flexible manufacturing systems and other computer-assisted functions, into a unified system.

Computer integrated manufacturing (CIM) has three basic components:

- The advanced computer controlled manufacturing technologies which are becoming commonplace in today's manufacturing environment.
- The computer based design, planning, scheduling and control applications which many companies are also adopting.
- The interconnection between these components to allow information to be transferred automatically between them.

The essential characteristic that differentiates CIM from individual automation projects is information.

This book is about information strategies. It sets out to

give practical guidance for the development and implementation of a company information strategy, which for many companies today may be the singlemost important guarantee of their future profitability and even viability.

In a world where advanced manufacturing technology is equally available to all the industrial nations, it is those companies whose managements are best informed and most able to respond quickly to fast-changing market demands which are going to survive and thrive. Such capabilities are difficult to achieve, and management 'flair' is not an adequate substitute. They are the fruit of a carefully structured strategy for information, implemented right across the company over a period of years.

A development which is simplifying the possibility of implementing a company information strategy in a manufacturing environment, and bringing it within the range of financial viability is the trend towards the adoption by leading users and vendors of computer systems of a set of international standards for communications within manufacturing company environments. For industrial applications these standards are grouped under two specifications: the Manufacturing Automation Protocol (MAP) and the Technical and Office Protocols (TOP). Each has reached its present level of acceptance because of the determined sponsorship of a major US company – General Motors in the case of MAP and Boeing for TOP. One must also give credit to the vision and enthusiasm of one man, Michael J. Kaminski, head of the General Motors MAP programme, who has been the principal dynamo driving forward this whole standardisation activity in manufacturing communications.

MAP and TOP are making possible a new level of integration in advanced manufacturing technology. In themselves, however, they are no more than the plumbing that makes a central heating system possible; the electric wiring that allows you to plug in a coffee maker or a microcomputer. What MAP and TOP have done is to remove some of the financial and technical barriers to the implementation of a company information strategy. Because MAP and TOP are relatively new technologies, and because it is important to be aware of them when considering how to go about implementing CIM, this book considers their place in the context of a company information strategy, and contains basic information about them.

The book has been written in close collaboration with the Manufacturing Communications Group of Coopers & Lybrand Associates who, in addition to their extensive experience as management consultants, now have to their credit the project management of the CIMAP event already mentioned. The fact that such a complex demonstration, involving fiercely competing companies, could be mounted so successfully in only nine months, is of great credit to the Coopers & Lybrand team, as well as to the goodwill and hard work of the participants and the robustness of the MAP and TOP specifications.

Jack Hollingum
February 1987

1 THE NEED FOR BUSINESS INTEGRATION

Information is the most valuable resource in industry today. It provides the basis for the effective management of a business

This book is founded on the belief that information is the most valuable resource which a manufacturing company possesses today. That is not to say that it is the most costly resource – though for some companies it may be. What is meant is that a soundly based and well implemented strategy for information can give a company a lead over its competition on a scale which is no longer possible simply by efficient use of materials and labour, or even by adopting the latest production technology.

However, the availability of information has this value only if it allows management to use it effectively in the management of the business. Information must therefore be considered not only in terms of its content but primarily in terms of its: presentation, accuracy, timeliness, and completeness. Integration of the information available within the company is essential to address these issues and provide a basis for real and effective management.

Advanced production technology is equally available to all competitors in world markets and can no longer be relied on as a significant source of competitive advantage

How has information come to assume critical importance for manufacturing industry? One answer is to be found in the fact that the technologies on which companies used to rely to give them an advantage in international markets have largely become international property. High quality machine tools, assembly machines, robots and other advanced production technologies are equally available all over the world, in low-wage economies as much as in high-wage economies (Fig. 1.1). Computers and business applications software are equally widely available, and companies everywhere are keenly aware of the need to keep costs down by efficient use of materials and labour.

Another answer lies in the changing market pressures facing manufacturing companies today. We are moving from an era where the focal point of competition was product cost to one in which companies must consider a much wider range of pressures.

- Cost is still a key factor, but the balance of cost is shifting as the overhead content constantly increases.
- Customer service, in terms of reliable delivery and short lead time, is increasing pressure on planning and control of operations.
- The market is expecting a much wider product range, requiring manufacturing flexibility.
- The acceleration of technology increases the pressure to get new products onto the marketplace quickly.

Information provides a key with which management can effectively address these conflicting pressures, and is therefore a significant means of gaining a competitive advantage.

The consequence for each individual company of the increasing pressures in today's international free market is that one way in which it can gain a clear advantage over its

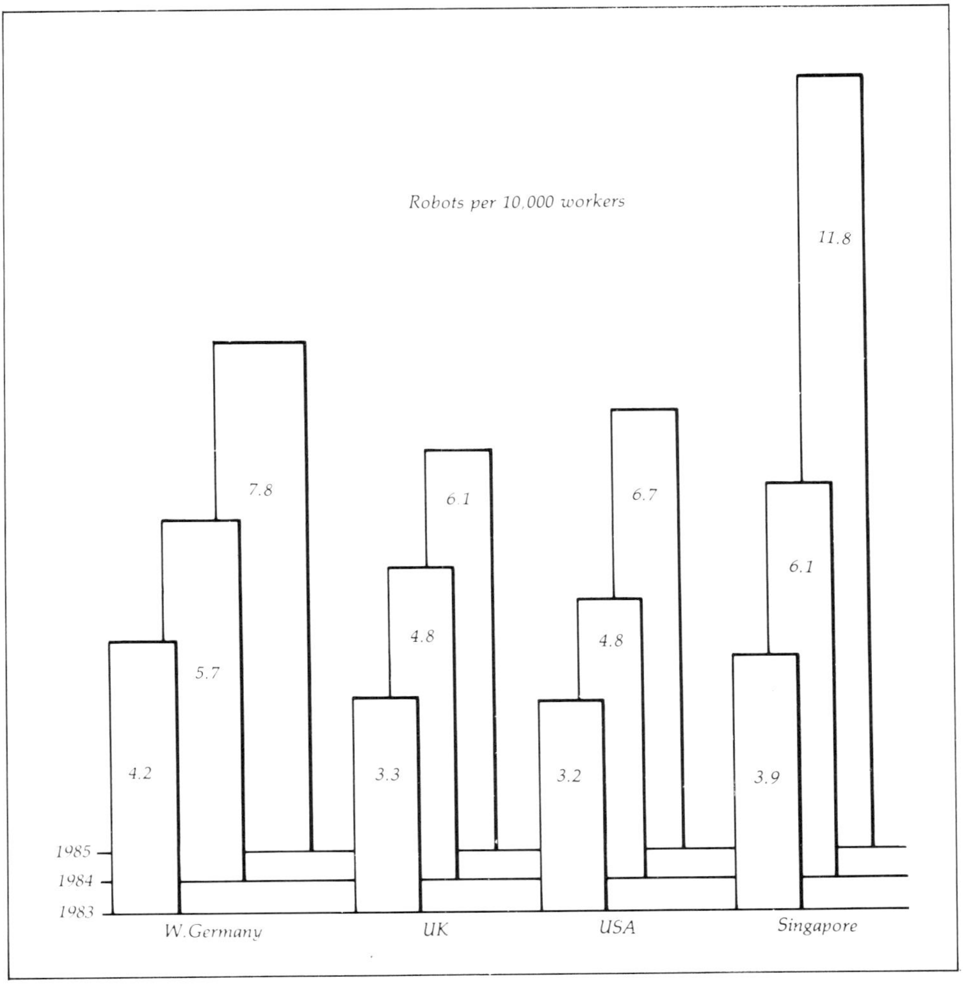

Fig.1.1 Distribution of robots related to working population for three developed countries and one developing country

competitors is through the skill of its management in making proper judgements based on reliable and timely information. Any gaps or inaccuracies in the information on which the top people in a company – and not only the top people – make their key decisions can have major consequences for the company's profitability and even its viability.

It is becoming increasingly apparent that management is the most significant competitive weapon in world markets: information is the foundation upon which sound management is built

This is not just a matter of providing a sound basis for business 'hunches'. There are many ways in which the efficient use of information can give management the freedom to win a strong competitive advantage. A company which can halve the time required to get a new product from the drawing board into production, through the effective use of information technology, will frequently be able to obtain a commanding lead in the marketplace.

At every level concerning business efficiency, information must be seen as a business asset, a 'factor of production' (Fig. 1.2) which companies will have to manage much more efficiently than they do today in order to get the benefits from it. Most companies are well aware that people are important and represent an asset that must be managed efficiently. Companies also appreciate the

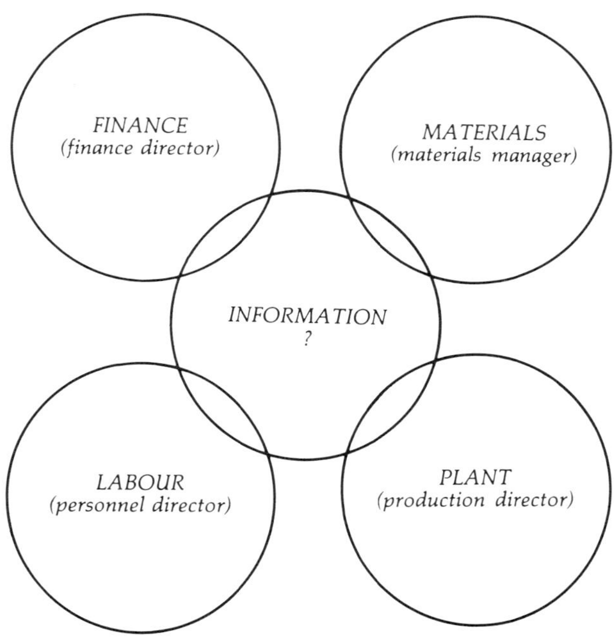

Fig. 1.2. Who manages these resources in your company?

need to use materials and other resources efficiently and to keep abreast of the latest developments in technology. Not many, however, recognise the cost and significance of information as a business asset.

The management of information often accounts for over 2% of sales turnover. This expenditure must not be made in a piecemeal fashion

It is not often appreciated how large an item information technology has become in the costs of manufacturing industry. It has been estimated that between 2% and 4% of industrial sales turnover today is spent on computer related technology. In other words, on average, that proportion of the price of consumer goods in the High Street and of the capital goods purchased for factories is going into computers, software and all the related expenses such as maintenance and training. Information technology is becoming a significant proportion of industrial costs.

Information is more costly than most people realise. Computers can help reduce its expense, but they can also increase the risks

In many cases these costs are incurred purely in the collection and maintenance of information. Many companies have realised its importance and have set about the task of gathering and hoarding information, often at very high cost. But the value of information lies in its effectiveness as a tool for decision making. Very few companies recognise the additional costs incurred in transforming the raw information available in the computers into usable management information. These costs are very often clerical and give rise to loss of accuracy and timeliness, and are very often associated with the gathering and collation of information from different functions within the company.

The increasingly widespread use of computers in industry (Fig. 1.3) is as much a risk as an opportunity. Where misapplied, computers become a source of high costs, large overheads, and great inefficiency, which do not service the true objectives of maintaining that information, namely those relating to the management of the business.

Computers have become widely used in all areas of industry and are embedded in very many pieces of production equipment. It must be understood that the problems and opportunities that arise from the use of computers extend right down to the shop floor. Machine tools, robots, material handling devices, sensors and

Fig. 1.3. Types of microelectronics equipment used in British industry, based on a sample of 1,200 establishments (from a research report by the Policy Studies Unit)*

	1983	1985	1987 (expected)
BASE	1200	1200	1200
TYPE OF EQUIPMENT USED			
CAD workstations	10	17	29
CNC machine tools	18	24	27
PLCs	24	43	45
Machine controllers	13	19	22
Process controllers	14	20	
Machine controllers or process controllers or both	21	29	
Pick-and-place machines	15	9	17
Robots	3	7	12
Pick-and-place machines or robots or both	7	12	22
None of the above types of equipment	14	9	4

*Percentages of all establishments in sample using each type of equipment

As information becomes available from the factory, the collection and use of it must be carefully planned and structured

controllers often have computers incorporated into them. An information strategy for a manufacturing company must encompass all these 'intelligent devices'. The opportunity to collect, analyse, combine, consolidate and present information from all these devices is the essence of CIM.

There is a danger, as technology becomes available

which makes CIM technically more feasible, that companies may rush into expensive purchases before analysing carefully what their real long-term requirements are. The result is liable to be a piecemeal approach which leads to:

- Missing data from some areas.
- Duplicated and inaccurate data.
- Complex systems and procedures.
- Little flexibility.
- Unmanageable information.

The implementation of computer systems in such a piecemeal manner often covers up the real problems and institutionalises inefficient working practices. The result is: increased overheads, organisational complexity, and waste.

Many businesses are currently experiencing the results of such a piecemeal approach but perhaps do not recognise the same mistakes being made in other areas of the company. The effects of a piecemeal approach can be seen in many of the 'home-grown' MRP systems in industry, which grew as a result of tackling each new problem as it arose. Nobody took a broad view of the needs, and the resultant systems institutionalised unsatisfactory working practices in many different departments, were surrounded by an impenetrable aura of 'mystique' (Fig. 1.4), and most importantly did not serve the needs of management.

By contrast, the professional, considered approach to information management is characterised by many modern MRP-II systems:

- Simplicity of design.
- Ownership of data.
- Accuracy of data.
- Co-ordination of departmental activities.
- Flexibility.

The implementation of systems in a professional manner as part of a strategic plan, as in Fig. 1.5, allows management to highlight inefficiencies and the needs for organisational change. Such an approach results in:

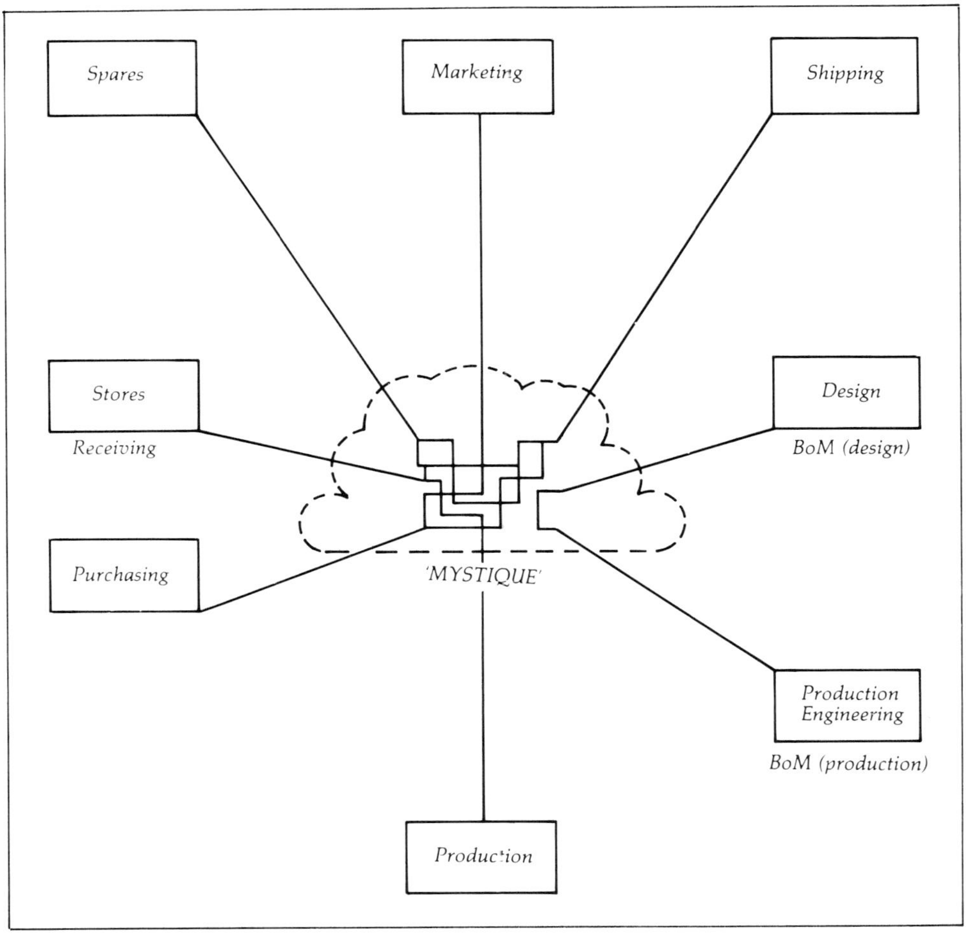

Fig. 1.4. A piecemeal approach to MRP systems in the past has created confusion and mystique. Invariably such an approach has been dramatically unsuccessful

organisational improvements, and rapid payback and high return on investment.

The lessons learned from these historical experiences in manufacturing companies *must* be carried forward by companies now considering CIM, as the cost and risk of failure are going to be much higher.

Information is vitally important to modern business. It is

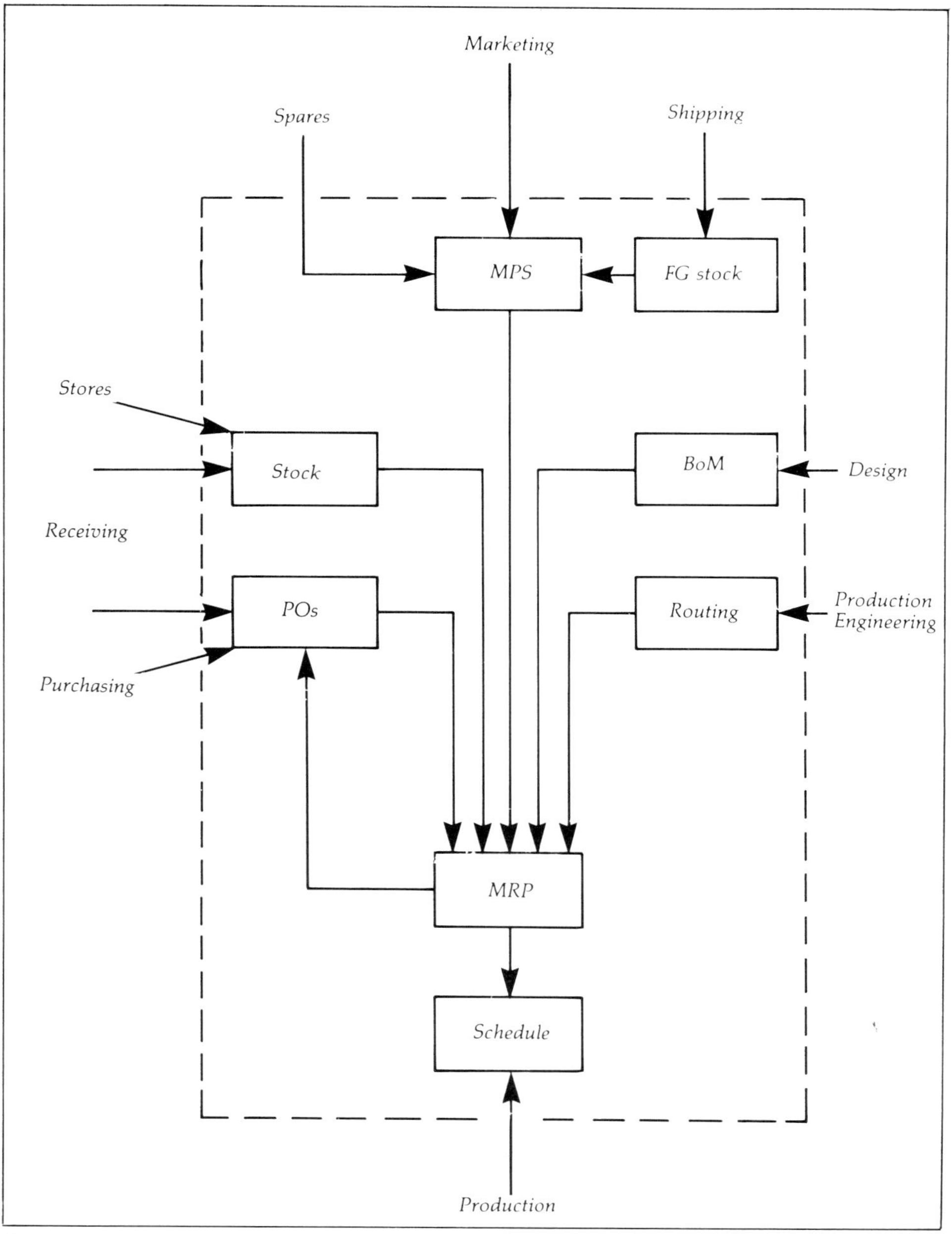

Fig. 1.5. A consistent and systematic approach to MRP resolves problems and speeds payback

costly and must be used efficiently. Its management should not be left to chance but should be in accordance with a carefully defined strategy. Manufacturing industry's track record in information management is not good, and as the opportunity and competitive pressures to move towards computer integrated manufacturing mount, we must think strategically about what we want information for and how we will use it to manage our business. The level of investment predicted in factory information indicates that the costs of failure are going to be high.

The next chapter will take a closer look at what an information strategy consists of and how it can be developed.

2 MANAGEMENT OF FACTORY INFORMATION

Every effort must be made to avoid institutionalising inefficient working practices when implementing information technology

Factory management is often described as a balance of the four M's:

- Men.
- Money.
- Machines.
- Materials.

In order to manage these resources effectively, managers require timely and efficient factory information. Traditionally this factory information was provided by telephones, letters and face-to-face communication. But today, computer and communication technologies have developed which are very much more efficient in the way they can collect, analyse and present factory information.

To be effective these new tools need to be applied at the correct point in the management process. There is no benefit to be gained from automating bad management practice. The first step to effective use of these new tools is to understand the role of factory information and the principles by which it can be utilised to provide more effective control (Fig. 2.1).

Information is a resource, just like materials and labour. It is costly to generate, and costly to use

It is necessary to appreciate that factory information is a resource just as much as materials, men and money, in fact today it is a particularly expensive resource. It is expensive to generate in the first place, and additional cost is incurred every time it is transferred, copied, worked on or revised. For example, product data, such as a design drawing or bill of material, incurs initial costs on account of the large amounts of time a designer or design team invested in its creation. Throughout the life of the product, additional costs are incurred as product changes are incorporated into the product information, and as the information is disseminated to users of it, by whatever means.

Take a look at the information sitting on your desk. Think what it must have cost to get it there, consider the time invested by highly paid people and the multitude of routine activities like typing, printing, editing, transmission, which the information has gone through.

Factory information is a business asset. It has a measurable cost of production, transmission and management. Computing technologies allow us to reduce these costs, and indeed the cost reductions are very often the basis for justifying computerisation.

Information is of little value unless it is being used

In contrast to all other business assets, however, which depreciate in value as they are used, information is valueless unless it is being used. The more it is used, the

- *is it essential for decision making?*
- *is its purpose clearly understood?*
- *is the company organised to make best use of it?*
- *is the essential information concealed in irrelevant data?*

Fig. 2.1. Key questions to ask all information users

more its cost of creation can be justified (Fig. 2.2). One could think of the cost of creating a unit of information being recovered through the decisions made upon it. Product information is clearly easier to justify than perhaps a local accountant's variance report detailing discrepancies in labour content for a low-value manufacturing process.

Information is expensive and its use should be justified, as one would for any resource such as materials and labour

Information costs should be justified. If you are in any doubt about this consider for a moment what is likely to happen if no justification process is performed; assume for a moment that the resource is free. The result is that, as with most free commodities, it will be over-consumed. Information will be generated which is of little value, and will be duplicated even when it is obsolete. This sort of

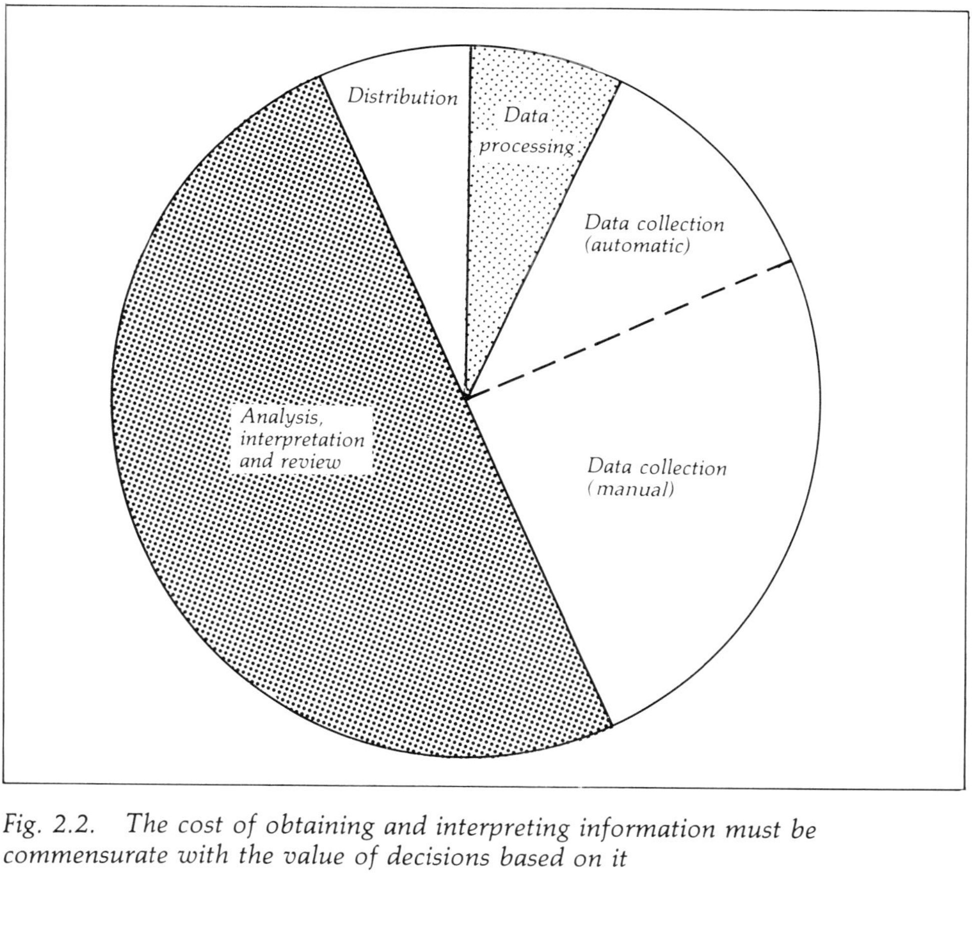

Fig. 2.2. The cost of obtaining and interpreting information must be commensurate with the value of decisions based on it

redundancy in information is wasteful and counterproductive, it actually hinders the efficient management of valuable factory information. Because, however, the use of computing technology loses the cost of information in general overheads, this is a pattern which is very common in manufacturing environments. Since users of factory information do not see a cost associated with it, over-consumption is a common symptom, with a resulting degradation of the effectiveness of that information in serving management objectives. The cost of information must be set against the value of decisions that are taken on the basis of that information.

Factory information as a resource is something which is overlooked by most of us in our day to day management activities. It is therefore necessary to highlight a small number of simple factors that should be considered before any new technology is applied to managing factory information:

- Factory information is a costly item in any business and should be recognised and managed accordingly.
- Generally, the most valuable information is that which contributes directly to management decisions.
- The function of accurate and timely information is to maintain the quality of management decisions aimed at controlling other key business resources (such as the four M's).
- The effectiveness of factory information is based upon quality, not quantity. It should be possible to make a decision without the need to gather enormous amounts of supporting information.

Establish working principles to utilise factory information

Once the above factors are realised within a organisation and it is acknowledged that there is a need to manage information more effectively, it is useful to establish working principles aimed at making effective use of factory information. Only when these principles are established should we consider the alternatives for generating this information.

- Never ask for factory information unless it is specifically needed to aid a decision. The collection of such information is a waste of a costly resource unless applied immediately to a justifiable decision.

- Realise that although information is often computer generated the effective use of it requires people – people who clearly understand the business objectives, and who appreciate the significance of factory information in terms of those goals.
- 'Factory workers' can improve their productivity through automation and computers; 'information workers' need the same technological aids to ensure their decision making is equally productive. However, the best computers and communication systems in the world will not overcome poor management practice.
- A surfeit of information is counterproductive. Information must be presented at the level of detail where it contributes effectively to supporting the good judgement of business managers. Information for its own sake does not support management decisions. (How much paperwork do you have on your desk today?)
- The golden rule when collecting or generating factory information is to ask the question: 'Do I need this information and what is the cost of providing it?'

When considering a migration towards CIM, these principles of factory information must not be forgotten. It is all to easy to launch into the introduction of computerisation on the factory floor, and integrated communications, without considering the real value of the information that our investment allows us to collect. If we do this, we may well achieve the local improvements in productivity that we would hope for, but may miss the opportunity to improve our ability to manage more effectively using the high quality, relevant information which CIM can provide.

3 A PRACTICAL WAY TO BEGIN

Information technology principles must be confronted with the real limitations faced by the company

The rest of this book will talk about the practical steps that will have to be taken in developing an information strategy and a systems architecture, and in following the strategy through to implementation, in order to ensure that the large investments made in CIM will support an overall strategy for the use of business information. The approach will be founded on information technology principles which are proving effective in companies leading the way in computer integrated manufacturing.

The focus of the book is on how a business can apply manufacturing and information technologies to gain improvements in operational efficiency and management quality through the adoption of a strategy to manage and use its information resources to maximum effect. In a company adopting information technology for the first time on a greenfield site, the principles may well be put into practice without too much difficulty or modification.

Most businesses today, however, are faced with a situation where computers have been installed and applied, usually incrementally and with independent systems for different functions, over a long period of time, perhaps rather like Fig. 3.1. Such companies already have a large investment in computer hardware and software, and in experience of using them – though there is often little communication between these separate 'islands of information'.

The other major difficulty is that operating methods and the ways of using information have become entrenched with the passage of time – often to the extent that a company is no longer operating in a way which is consistent with its basic objectives. It may even have lost sight of its objectives and be operating primarily in order to serve the information systems created in the past.

For example, companies often maintain a vast amount of product information which was originally designed to service systems and requirements which may no longer exist or which have evolved as a result of changes in the business environment.

How to get out of the rut of current thinking

If we can accept intellectually the argument that we should be developing an information strategy driven by our business objectives, then we are faced with the question of how to 'unfreeze' current thinking and redirect it towards redefining how the business ought to be operating. This fresh, unblinkered view is essential before we can take the next step of planning the migration to a more appropriate architecture which fully supports working methods derived from the business objectives.

The process must be one which effectively frees thinking from current constraints at all levels of the business. An effective means of achieving this 'unfreezing' is to conduct a critical review – a diagnostic – of the current situation, and then to acknowledge and agree upon the shortcomings

Computer aided design

Data centre

Machining cell

Fig. 3.1. Most businesses have acquired independent computer systems for different functions over a period of time, resulting in 'islands of automation'

highlighted by this process, at all levels in the organisation. Some headings for such a review are shown in Fig. 3.2.

Make unkind – but fair – comparisons with the best in your industry

To maximise the effectiveness of such a review, the company must conduct an unkind, but fair, comparison of its own internal operations against a standard set by what is seen to be the best achievable in the market in which the business operates. This is an area where the use of independent consultants can be of great value – because of their independence, as well as their experience of the standards that others are setting, and also because internal staff may be inhibited from being too critical of existing practices.

The objective is to awaken the company at all levels to the fact that current practices, systems and technology may not be coordinated with business objectives, and to identify possible areas of risk and opportunity for improvements.

Be positive in criticism – look for opportunities

One of the most important features of the diagnostic is that of maintaining a positive perspective. The aim is not to castigate departments and functions for past failings but to free the organisation from its past and identify the opportunities for future improvement.

The diagnostic comprises a detailed review, function by function, encompassing the following points:

Engineering and Technology policy
Engineering Data Management
New Product Engineering
Production Engineering
Manufacturing Technology
Quality Assurance
Finance
Marketing
Sales
Logistics

Fig. 3.2. A broad ranging study is often needed to stimulate fresh ideas and develop a collective vision of the future

- Existing operational procedures.
- The use of systems to support them.
- The effectiveness of computer technology.
- The availability and quality of information.

By getting all the business functions to recognise and own these weaknesses, as well as identify the strengths of the business, the foundation is laid for the positive mobilisation of an information requirements development plan, based on a wide recognition and understanding of the need for change and a common perception of the broad outlines of a plan for change.

Use the diagnostic as a reference point

Such a diagnostic also provides a valuable reference point in three respects:

- It crystallises what are the fundamental strengths and weaknesses of current operations.
- It represents a documentary record of today's position that can be used to confirm that any future strategy does actually represent an improvement.
- It provides the starting point for measurement of improvements in performances.

4 DEFINING USER REQUIREMENTS

An effective information strategy must be derived directly from the objectives and strategy of the business

A strategy for information is not something to be considered in isolation from a company's overall business strategy. Its purpose is to contribute to the achievement of the company's business strategy. The business strategy must therefore be clearly stated before an effective information strategy can be developed.

The development of an overall business strategy is beyond the scope of this book but a brief discussion of the subject will help to show how important it is to state business objectives and strategy clearly before deciding on the information strategy.

Very often the best foundation on which to build a

business strategy is an international comparison of the company's strengths and weaknesses in relation to its competitors. Part of such a study may reveal comparative figures like those in Fig. 4.1, indicating serious weaknesses in a number of areas. Careful study of relative performance in each of these areas will lead to the establishment of a long-term strategy, as well as priorities for action, upon which fundamental operational objectives can be based.

Many senior executives assume that the basic business objectives are obvious, so they do not realise the need to provide a clear statement to the next level of management responsible for the individual functions within the company. This frequently leads to functions working to objectives which seem obvious at their own functional level, but which are not necessarily in complete harmony with the basic objectives of the business. The business strategy must be stated in terms of key objectives, policies, operating characteristics and constraints in order to be an effective driving force for the operation of the company. For example:

- An objective to achieve market share is admirable, but does not provide any direction to the functions within the business.

Fig. 4.1. Compare your company's strengths and weaknesses with international competition

	Company 'X' (UK)	*Company 'Y' (Japan)*
Stock turn ratio	*4*	*15*
Indirect/direct staff	*1.4*	*0.5*
Sales per employee	*£30k*	*£90k*
Development lead-time	*100%*	*70%*
Manufacturing lead-time	*100%*	*50%*
Product cost	*100%*	*70%*
On-time delivery	*75%*	*95%*

- The strategy must state how this objective will be achieved:
  - through rapid introduction of new products? If so, which product ranges, what lead times are being sought, and so on?
  - through manufacturing flexibility? If so, how? Using automation?
  - through short lead times? If so, what are the market lead-time requirements?
  - through higher quality? What specific quality objectives is the business seeking?
  - through low cost? If so, where are the cost targets, which are the sensitive product ranges?

The business strategy must be used as the basis for the production of separate but coordinated strategy documents covering, say, financial control, marketing, product engineering and manufacturing, as in Fig. 4.2. These

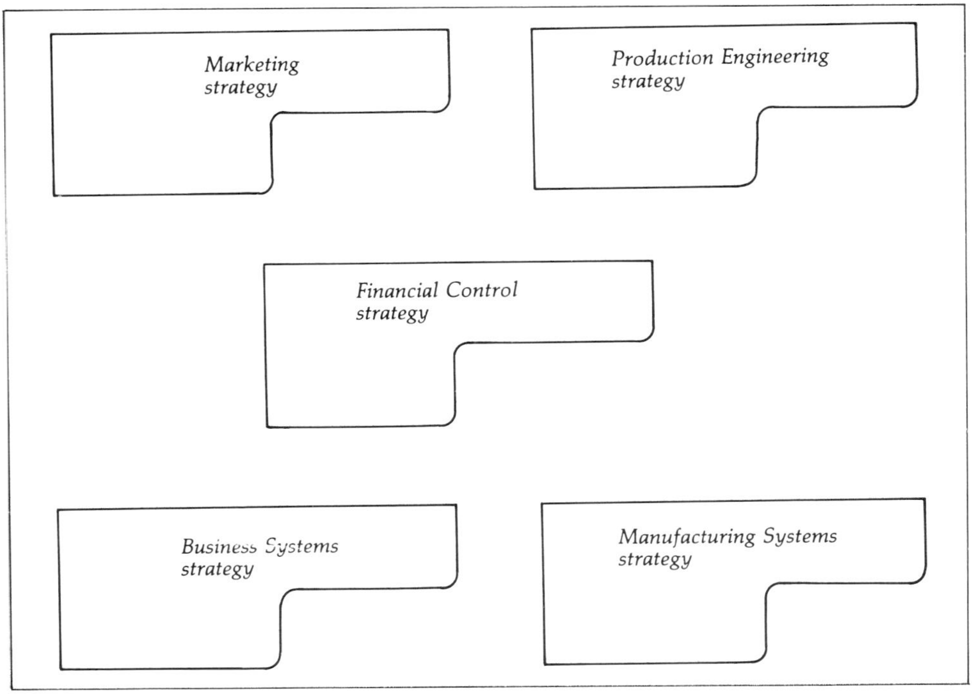

Fig. 4.2. Business strategy must be reflected in a set of coordinated documents

strategies must be the foundation upon which the information strategy is developed.

Define the functional requirements of users in the light of the coordinated objectives and strategy of the company and the functions within it

The first step in developing an information strategy is to define the functional requirements of users for information in the light of the stated objectives of the company and the functions within it (Fig. 4.3). This 'Requirements Definition' must be formally documented as the basis for proceeding with the subsequent stages which design and implement the strategy.

In order to produce a statement of its information requirements a company must define exactly how it is going to operate and the principles which will drive all its functional activities. It must reach agreement on how these activities will be carried out.

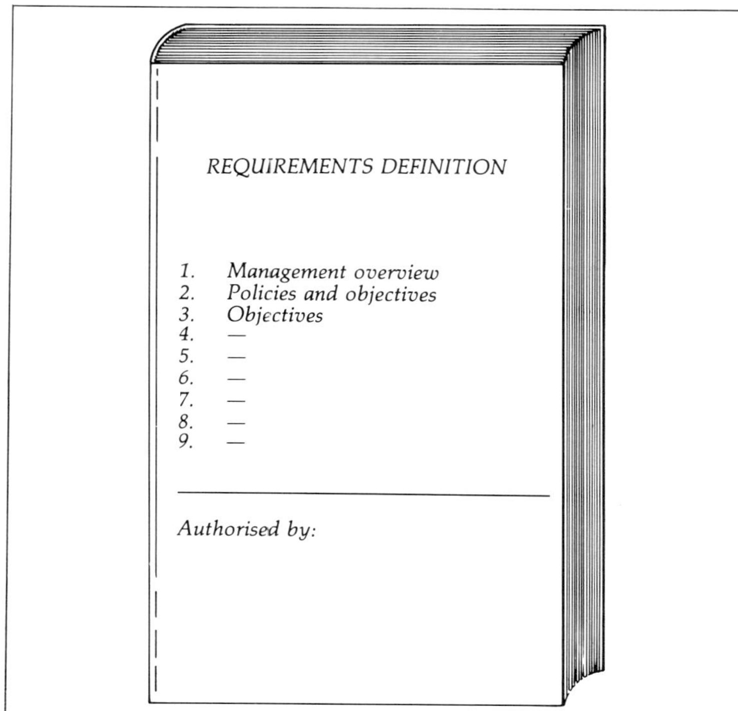

Fig. 4.3. A Requirements Definition must be formally documented, detailing the functional requirements for information

In developing the Requirements Definition many value judgements must be made by senior managers in the light of company objectives and strategy

The Requirements Definition must cover a time horizon of at least five years and possibly as many as ten. It is not a statement of what people think they need in order to support their activities. It is an agreement on what *should be provided* in order to operate consistently with the agreed objectives. The document is the result of many value judgements. The business as a whole must assess the relative importance of various activities and be able to identify where operating procedures need to be revised. Only experienced users can review the basic objectives and working methods of the functional departments and develop an agreed, coordinated set of operating principles.

For this reason it is important that the Requirements Definition stage should be undertaken not by information technology specialists but by an experienced team who are familiar with all aspects of the business and are able to question the existing methods of operation.

This usually means the formation of a multidisciplinary project team from within the middle management level of the organisation. A company must face up to the fact that a significant amount of input from these often critical management resources is necessary, and that senior management time must also be committed to the mobilisation and direction of the Requirements Definition project. These resources are frequently difficult to provide because of day to day operating pressures, but the need for them must be viewed in the light of the fact that the project team will be laying the foundation for the basic method of business operation for the next five to ten years, and are therefore defining the means by which the stated business strategy will be achieved.

The team responsible for developing the Requirements Definition must question existing working practices and avoid developing systems that hide problems and institutionalise inefficiencies

A large company may already have or be able to assemble such a team to carry out the Requirements Definition. Other companies will probably seek support from independent consultants who are able to bring relevant experience to bear. The danger to be avoided is that of handing over the entire responsibility of the project to outside consultants, and therefore losing the business ownership and identification with it.

The value that outside support can bring lies in the

awareness which can be added to the project of alternative working practices and technological opportunities. Although the Requirements Definition is largely an introspective view of how the business should operate, awareness of the tools available to support possible alternatives is an important feature of the information strategy phase.

A confident and determined approach is required when trying to deploy information technology to obtain strategic advantage: it is usually expensive and involves significant organisation change

In preparing a Requirements Definition for information strategy, the project team will have to talk openly to people about their jobs, their objectives, how well they can function with existing information, and what tools will bring most benefit in the future, in terms of focusing the way in which they work on the real business objectives. These interviews should uncover the main areas where information is required to support operational decisions, and therefore where it is of greatest value.

On the basis of the interviews the requirement for information can be defined, either as a requirement for communication – for example the communication of stress analysis information to a CAD system – or as a basic application requirement, for example the need to record quality data during inspection.

There should be no 'sacred cows' during the strategy phase. If the business organisation will need to change to meet business objectives, then the issue must be raised and addressed by the project team. There is no point in pretending that technology will resolve these problems – it will most probably exacerbate them.

The final Requirements Definition document is the main product of the strategy phase and must have the commitment of all levels of management

The Requirements Definition document is the main product of this stage of the project. It is not a 'technical' document: it does not discuss the technical aspects of how information technology should be applied. The document simply defines the functional requirements for systems that will support the activities of the business.

It is important to write this document in language that everybody can understand. It must be written well, distributed widely, and presented with enthusiasm, so that it forms an effective vehicle for motivation and consensus building – Fig. 4.4.

1.	*Management overview*
2.	*Buisness policies and objectives*
3.	*Functional objectives*
4.	*Functional principles, policies, and procedures*
5.	*Functional overview of business control system*
6.	*Information requirements*
7.	*Business benefits*
8.	*Outline implementation plans*

Fig. 4.4. Topic headings to be addressed in a Requirements Definition

The first part of the Requirements Definition report presents a management overview and clearly identifies the next steps in the process of developing and implementing an information strategy.

Part 2 seeks to position the reader with respect to the wider objectives and strategy of the business. This can just be a summary of the business strategy described in other documents, if these exist, but frequently companies have no statement of strategy. In such cases the development of this part of the Requirements Definition takes longer, but it can never be ignored.

The contents of part 3 are largely educational. It documents the objectives or 'charter' of each function within the business.

Part 4 discusses the principles, policies, and procedures which influence the functional activities of each department. For instance, this part might talk about functional policies to use in automation and operational principles such as dual sourcing, credit control, and so on.

Part 5 explains the how the activities of the various departments will be controlled and coordinated, in essence a description of the functional operation of the business control 'system'. For instance, this might talk about how the Master Production Schedule will drive the planning process, how design data will be used to generate production engineering information, how quality data will influence purchasing and inspection, and so on. It seeks to create a non-technical 'model' in the minds of everybody, to explain the operation of the business as a single system, and the flow of information around it.

Part 6 consists of a series of tables which set out the information required by each function of the business. These tables identify the inputs, outputs, and internal processing of information from a functional viewpoint.

Part 7 discusses and quantifies the opportunities for business benefits as a result of the improved operating practices embodied in the document.

Part 8 presents outline plans for the implementation of various projects which will move the company towards the identified objective. Detailed and final planning of these projects can only be agreed once a technical evaluation has been performed and the systems architecture study has been completed.

The Requirements Definition is an authoritative statement of intent and must represent the commitment of a very senior level within the company and be understood at all management levels

This document will be the authoritative basis on which information projects and purchases will be planned. It must be authorised by a senior group within the company, ideally the CIM Steering Committee if one has been created (Fig. 4.5). Only when the document has been signed off at all levels of the business should the design stage be commenced, as described in the following chapters.

Product development
Manufacturing
Marketing
Finance
Information processing
Logistics

Fig. 4.5. The Requirements Definition must be authorised by a steering group consisting of senior members of the company

5 DEVELOPING A SYSTEMS ARCHITECTURE

A Systems Architecture is necessary as a framework for the planned development of hardware and software to satisfy the agreed statement of requirements

Once a commonly agreed definition of requirements has been established, the development of an information strategy can move into its design phase, which begins with the definition of the company's Systems Architecture (Fig. 5.1). The issues surrounding the development of the Systems Architecture are complex. It is necessary to form a technical team to determine an architecture which best meets the stated requirements of the company, which is technologically feasible and is cost effective.

The term 'Systems Architecture' is being used more and more today. The idea which it describes is not complicated but it is important to understand it. A Systems Architecture is a representation of the relationships that

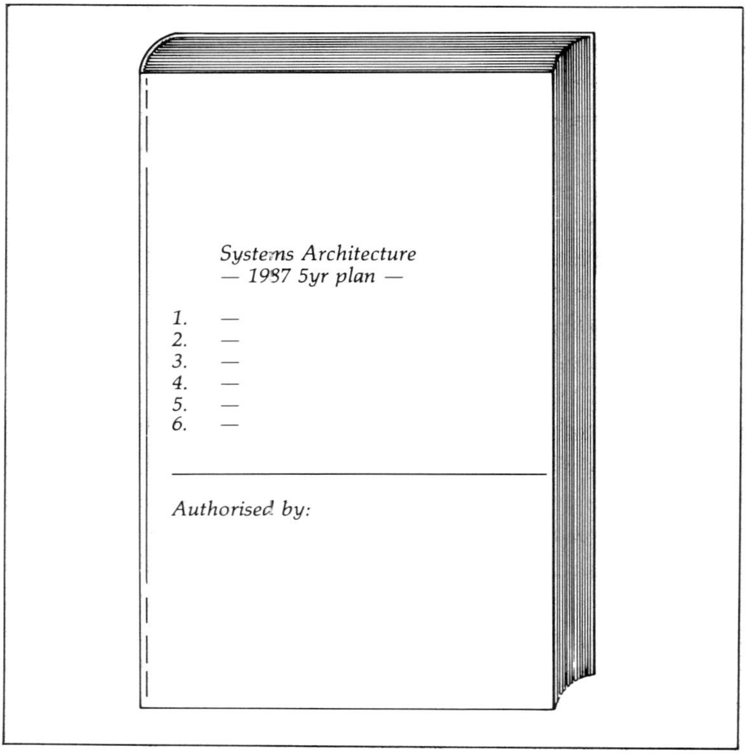

Fig. 5.1. The Systems Architecture document spells out in detail the company's present and projected systems and the migration path

exist between hardware and software and the information which flows between individual applications. It could be described as analogous to architect's drawings which represent the relationship between the structure of the building and the rooms within the building.

The need for a Systems Architecture becomes apparent as soon as detailed plans for the implementation of specific systems are discussed. Although functional requirements may have been agreed upon, if an overall architecture has not been developed individual functional requirements would not naturally tend to be satisfied by applications of technology which could be linked into an overall business system. Typical issues which must be addressed are:

The Systems Architecture reflects an agreed position on broad issues which affect all systems development activities. Two of the most significant of these are networking and distributed processing

- Should systems be distributed over several computers or is it better to develop them all around a central 'mainframe'?
- What sort of communications capability should be provided to support the development and integration of individual applications?
- Should a company-wide communications network be developed and if so what sort should it be?
- What type of computer equipment is required to operate the commercial software systems required by the company?
- How should the software systems themselves be integrated?

It is important to recognise that only the broadest issues that affect systems development activity need to be reviewed and decided upon at this stage. Any issues which can be left until specific implementations enter their detailed design phase should be postponed.

A technical team needs to be created to discuss issues that relate to Systems Architecture and agree a course of action. But users must also be involved

A very different team of professionals is required to develop the Systems Architecture from that needed to carry out the requirements definition (Fig. 5.2). The team needs to look in detail at the company's present manufacturing technology application programs and networking facilities, and at the changes that will have to be made in order to support new requirements. The team must obviously include specialists in manufacturing technology, data processing, and communications, but during this stage it is equally vital that the user team which developed requirements is involved in the process. Users must be involved in the assessment of applications software and made ultimately responsible for its selection and implementation.

The first step is to develop a picture of the Systems Architecture as it exists today

The first step is to form the team to develop a simple statement of the Systems Architecture as it exists today. This will include diagrams showing how the different pieces of hardware and software are connected together and the relationships that exist between them. What communication protocols are being used? What is the communications capability of the various pieces of computer equipment? Which computer runs which

Data processing *Systems design* *Networking* *Distributed processing* *Process control* *Manufacturing technology* *Application specialists (CAD, MRP,...)* *Trends in the equipment market*

Fig. 5.2. The Systems Architecture group require specific expertise in many areas

systems? What is the relationship between the various systems? All such questions need to be investigated and answered. The result is what Fig. 5.3 represents in greatly simplified form as the 'Systems Architecture today'.

The second step is to develop a picture of the Systems Architecture as it should be in five years time

The second step is to prepare a precise description of the systems architecture that is desired by the end of the five-year plan (Fig. 5.4). Several considerations will have to be taken into account in preparing the description:

- What manufacturing technology will we have when the plan is completed? This information should have been spelled out in the company's business and manufacturing strategy objectives.
- What applications will there be? This will have been set out in the requirements definition.
- Where will the applications run?
- What information should be transferred between applications?
- What is the frequency of information transfer? Are we talking in terms of milliseconds, minutes, shifts or days?
- What flexibility needs to be built in for the future?
- In the light of the previous considerations, what type of networking and communications will be needed? This is where an understanding is required of the technologies available and their characteristics.

Except in the rare event of a greenfield situation, account will have to be taken of the existing applications and

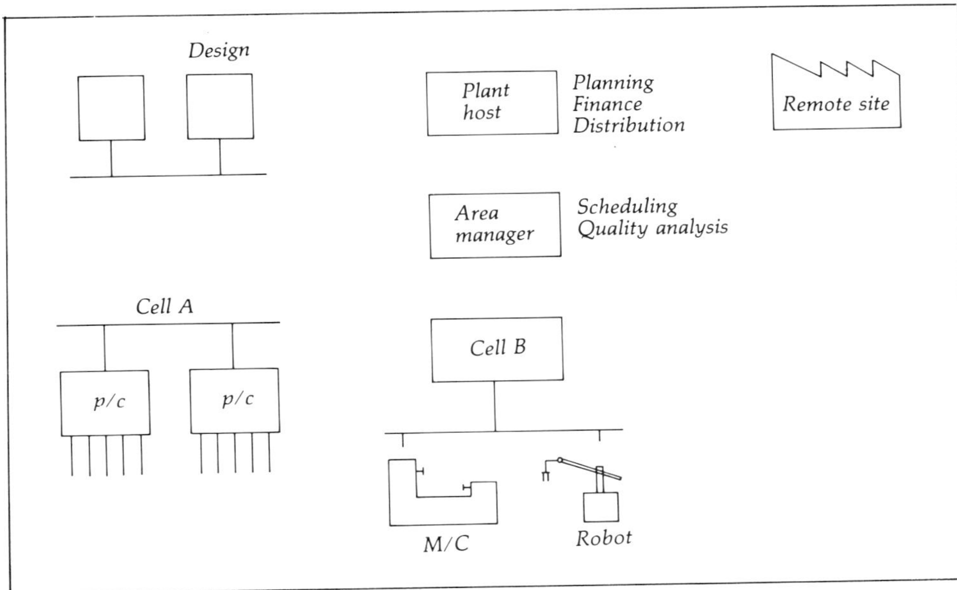

Fig. 5.3. A picture must be prepared of the Systems Architecture as it is today

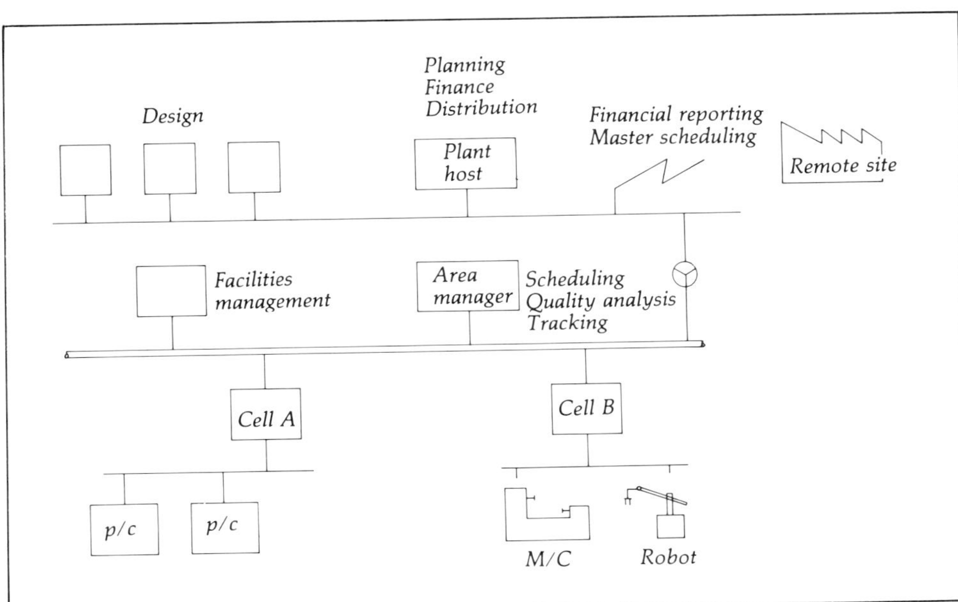

Fig. 5.4. A precise description of the Systems Architecture desired in five years time must be agreed upon

technology in the company, and these will strongly influence the transition to the new system.

Developing a Systems Architecture description requires a thorough technical understanding and a clear vision of likely developments in the marketplace. It is at this stage that decisions have to be taken regarding the relative merits of MAP, TOP, DECnet, SNA, Advancenet, and other proprietary communications networks.

A knowledge of developments in applications programs and the ease with which they can be made to communicate is also important. There are many standards being developed that address problems of exchanging data between systems. Standards such as IGES (International Graphics Exchange Specification) fall into this category and may need to be reviewed. A decision to standardise on IGES for graphics exchange will influence the selection of hardware and software.

The third step is the most difficult – to agree a 'migration plan'. This is a series of practical steps which will move the company from today's architecture to the desired architecture of the future

The third step is perhaps the most difficult. A plan must be developed to migrate from today's architecture to the architecture envisaged at the completion of the plan. This 'migration plan' (Fig. 5.5) needs to be a practical step-by-step programme for implementation of the new architecture.

The migration plan needs to be analysed into a number of manageable projects covering:

- Procurement and installation of hardware and software, and training in the technology.
- Development of the applications.
- Installation of the communications network.
- Transfer of applications to different computers.
- Development of interfaces.
- Implementation of stages:
  - installation
  - procedures
  - education and training
  - consolidation

The migration plan has to take account of interconnections and efficiencies of movement. During the

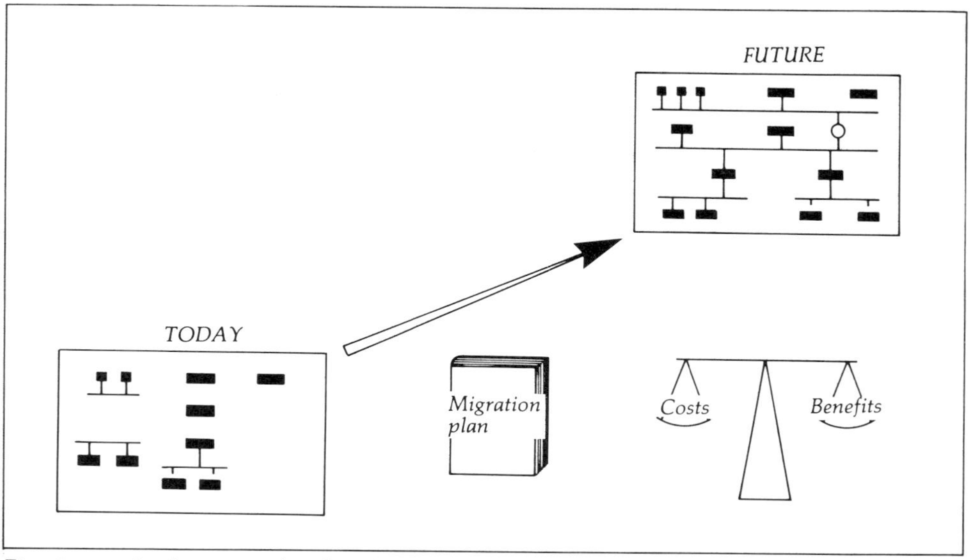

Fig. 5.5. A plan must be developed to migrate towards the desired architecture

migration process, interim solutions are inevitable, but disruptions caused by these must be minimised.

The plan should be structured hierarchically (Fig. 5.6). Within major phases such as shop-floor automation, and materials planning and control, there need to be individual projects such as process control of a production line and works order management. Individual projects should be further broken down into tasks such as system design, development and training.

The migration plan reflects agreements regarding the relative priorities within the systems development activity

It is not sufficient to agree on *how* the company will move from its current architecture towards its vision of the future. It is also necessary to determine *when*. Which activities should take priority over which other activities? How are the various activities dependent on one another?

Although the requirements definition will have identified priorities, it is only during the development of the migration plan that a firm schedule of activities can be defined. Factors which influence the relative priority of systems development activity include:

- *Short-term benefits from activities.* There are usually

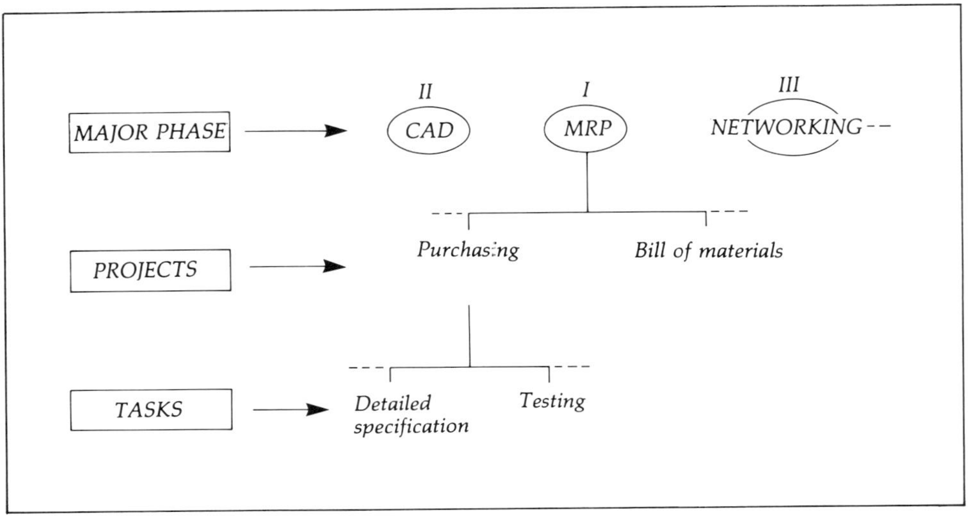

Fig. 5.6. The migration plan should be structured hierarchically into phases, projects and tasks

projects which offer immediate benefits for relatively small cost. These short term improvement projects should be started immediately.

- *Strategic significance of the activity to the company.* A project to provide the company with improved CAD functions will probably have a higher priority in this respect than a project to improve time and attendance recording.
- *Technical dependency of one activity on another.* Most projects are dependent to some extent on other activities. Most projects will be influenced by the provision of a company-wide network, so installation of the network may be one of the first tasks considered.

However, this is not the only consideration. Most of the activities reflected in the migration plan will have organisational implications. In the experience of Coopers & Lybrand it is these organisational changes that are the most difficult to accomplish, yet they also offer the greatest benefits since it is usually through organisational efficiency that integration activities are justified.

Technical issues and 'principles of design' both influence the development of the Systems Architecture

It is important to recognise that while there are many technical issues that influence the design of a Systems Architecture, there are also a number of factors that can best be described as 'principles of design'. Technical issues which frequently influence the development of a Systems Architecture include:

- Existing investment in equipment and systems.
- Availability of application software.
- Price/performance of hardware.
- Reliability of equipment.
- Strategy of equipment vendors.
- Availability of common communications capability.
- Availability of operations software.

Factors best described as 'principles' which frequently influence the development of a Systems Architecture include:

- Single (or limited) vendor policies towards equipment purchase.
- Responsibilities for 'ownership' and management of data.
- Company requirements for the operator interface.
- Emphasis to be placed on flexibility.
- Considerations of security and privacy of information.

A Systems Architecture document is the main product delivered by the architecture study team. Fig. 5.7 shows a typical structure for this document.

1. *Objectives and responsibilities*
2. *Description of existing architecture*
3. *Trends in technology and the marketplace*
4. *Principles of design*
5. *Description of desired architecture*
6. *Agreed migration plan*
7. *Human resource requirements*
8. *Costs and cost justification*

Fig. 5.7. Topic headings for a Systems Architecture document

The results of designing the Systems Architecture must be documented clearly, comprehensively and concisely

The first part usually discusses the objectives of the document and its position with respect to other documents, most notably the 'statement of user requirements' discussed in Chapter 4. It clarifies how the Systems Architecture has been derived from the requirements of the business as specified in the statement of user requirements, and identifies where compromises have been necessary and how these affect the defined requirements.

Part 2 presents details concerning the existing Systems Architecture. It identifies existing systems, their function and any integration that exists. It also presents details about the existing investment in equipment such as communication capabilities and control requirements. It is often convenient to present this part under three subheadings: Applications, Production Equipment, and Computers and Communications.

Part 3 discusses technology and market trends so that these can be examined in the light of known developments during each annual review of the plan. It is an important part because it documents many of the assumptions which might otherwise go unrecognised and unchallenged.

Part 4 defines the principles which have been the basis for development of the Systems Architecture. For example it might be taken as a 'principle' that it should be possible to maintain production for a period of one hour even if all communication to a production cell is lost. Or it might be a 'principle' that particular categories of application software should be written in a particular language and operate with a particular operating system. This part also describes the principles of design upon which the Systems Architecture has been based.

Part 5 is presented in the same format as Part 2 but presents details concerning the proposed future Systems Architecture.

Part 6 discusses the implications of the proposed Systems Architecture and charts a course for the company to migrate from where it is today (Part 2) to the future architecture (Part 5).

Part 7 discusses factors that relate to the human

resources that are critical to the implementation of the architecture. Does the company face a shortage of particular skills? What must be done in the way of training and education to help people through periods of organisational change.

Part 8 presents the project costs and justification on which the proposed migration is based. As this is one of the most difficult aspects of computer integrated manufacturing, some principles are put forward in Chapter 9 of this book.

The final Systems Architecture document must be authorised at a senior level

The information strategy document will be the basis for the planning of information projects and technology purchases. It has significant long-term implications for company profitability and capital investment. It must be understood and authorised by a senior group within the company, most probably the CIM Steering Committee if one has been created.

Plans should be reviewed every year in the light of technical and market developments, any proposed modifications being widely discussed and subjected to the same levels of authorisation as the original document.

6 THE ROLE OF FACTORY NETWORKS

When considering computer integrated manufacturing it is inevitable that the systems architecture will comprise a large number of computer based devices running applications which must communicate with one another. Networks are one of the key elements of CIM, as without them integrated systems running across many devices are impossible.

Networks have a critical role to play in the implementation of CIM

Any company which is using computers in manufacturing or office functions is probably already using networks to some extent. There may be simple point-to-point links with a host computer, or a more sophisticated bus network connecting a number of terminals in a department. In a large organisation there may be several mainframe computers connected together, carrying out financial and accounting tasks.

In a factory there may well be automated manufacturing cells or process control systems based on proprietary, or even bespoke, communications networks. There are unlikely to be any major communications facilities in place linking commercial systems to the shop floor.

Within their own areas of application these networks function satisfactorily. The trouble starts when the systems architecture requires these 'islands of information' to be connected together – for example connecting computer aided design to production engineering or to material requirements planning or to connect a robot assembly line to a production scheduling system. If the two sets of application software came from the same supplier it is possible – though by no means certain – that they will use a common networking system. It is not often, however, that a single vendor can offer a suite of application packages covering different functions, which will be seen by the user as the most suitable for each area of application. Almost inevitably the user will be seeking rapid implementation by purchasing 'off the shelf' CAD, MRP and process planning software and hardware, for example – and from different suppliers, using different networking methods, because he wants the best system for the job. The integration difficulties extend from the physical connectors and cables that are required right up to the problems of achieving application programs that operate correctly together.

To achieve communication, devices must not only be connected, they must also comprehend one another

Two sets of problems can be distinguished in this overall task of obtaining integrated information systems. The first group of problems is associated with achieving communication between the different systems in the company – actually getting messages from one computer or device to another. This area is the subject of this chapter and Chapter 7. The second group of problems is about what might be called achieving 'comprehension' between the systems. This is the subject of Chapter 8.

Until recently, the combined effect of these two groups of problems was to put up an almost insuperable barrier to the achievement of integrated systems. The widespread publicity now being given to MAP and TOP is well deserved because it signals the fact that the first group of problems is well on the way to being solved.

The Open Systems Interconnection model is the basis for future networking standards

MAP and TOP are the fruit of an effort which has been going on for a number of years to achieve what is called 'open systems' – that is, communications according to agreed rules, or 'protocols', in the public domain allowing equipment from any supplier to be connected successfully. The work was sponsored by the International Organisation for Standardisation (ISO) and a framework was defined, called the Open Systems Interconnection (OSI) reference model. This divides the communications task into seven layers, and the rules for the working of the different layers, and the interfaces between them, are precisely defined. The intention is that any change to a standard in one layer will not affect the others. A vivid depiction of the functions of the seven layers was recently given pictorially by George Turnbull of Eurotherm in his company's house magazine. A simplified version is reproduced here as Fig. 6.1.

The people most involved with the creation of the OSI reference model and with the associated standards activities were interested primarily in telecommunications and activities like electronic mail and financial services. They also chiefly represented the companies and organisations supplying the competing computers and services, and probably for this reason progress was slow.

In addition, the developing OSI standards were being driven chiefly by telecommunications requirements. They were not aimed primarily at the special problems of manufacturing communications such as:

- Systems distribution based on functional, not geographic, criteria.
- Widely varying data flows, ranging from large, infrequent file transfers in the technical office to short frequent messages on the shop floor.
- High population of connected devices, imposing a heavy potential load on the network.
- Hostile operating environment.

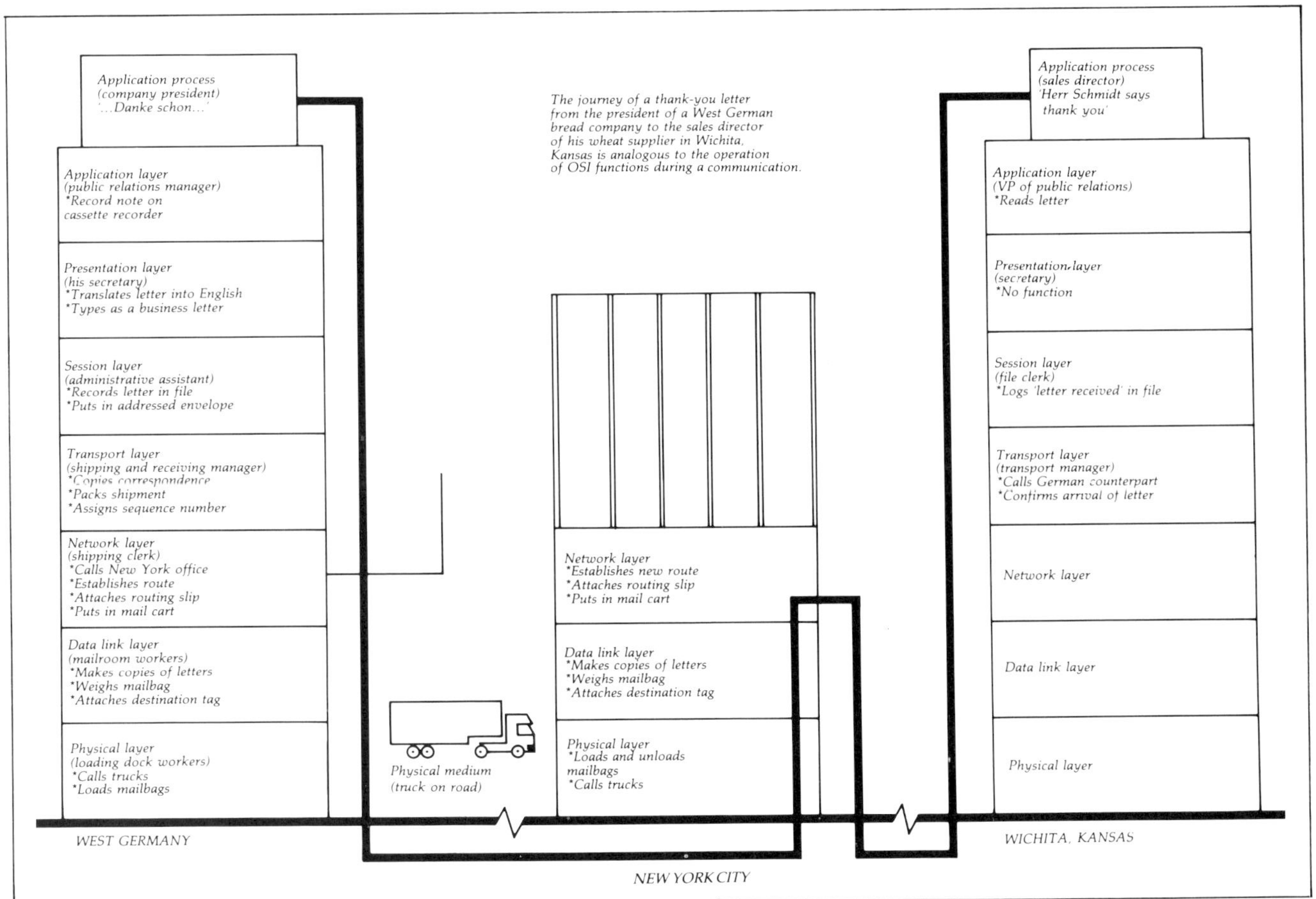

Fig. 6.1. How the OSI seven-layer structure connects two 'application processes', with a 'bridge' between them at New York City

The present acceleration of activity and the new emphasis on local factory communications is attributable almost entirely to one company – General Motors in the USA – which was facing major difficulties and prohibitive expense preventing the achievement of computer integrated manufacturing in its factories.

Factory networking standardisation pioneered by General Motors in MAP, backed by major users

A General Motors task force under the leadership of Michael A. Kaminski was given the job of identifying a set of standards which would be suitable for factory level communications. They quickly decided that OSI presented a suitable framework, and went on to select within it the standards which would be most suitable for manufacturing requirements. Unfortunately, at that time most of the standards were only in draft form, and some were no more than recommendations to adopt a specification from a member country. Nevertheless the GM team drew up its Manufacturing Automation Protocol (MAP) based on such standards as existed and other specifications which were expected to reach international status. A 'protocol' in this context means a set of rules which define how communication shall take place.

Having drawn up its specification, GM used the weapon of its massive purchasing power to persuade or bully independent-minded vendors to develop products compatible with the MAP specification. The company was also instrumental in the setting up of user groups in the main industrial regions of the world, and many other major companies are now backing the MAP programme.

MAP is described in greater detail in Appendix A at the end of this book, but it should be said here that the specification is evolving as standards become established, though 'migration' rules have been established to ensure upward compatibility of equipment.

Manufacturing office networking – sponsored by Boeing in TOP

The story of the Technical and Office Protocols (TOP) is similar but very much shorter. In this case aerospace manufacturer Boeing was the sponsor, and the first TOP specification was published in November 1985. It is intended to cover office functions in manufacturing companies and is identical to MAP in most respects except for the type of cable and topology which can be used at the physical communication level and the application services which are provided at the interface to application programs. At present the only such service specified – File Transfer Access and Management – is also included in the MAP specification, but other services are envisaged covering word processing, electronic mail, graphics interchange and other functions.

Why MAP and TOP are important

The claims of MAP and TOP to serious consideration in the development of a systems architecture have some strong foundations:

- They are rapidly becoming a widely supported standard, backed by many vendors and vendor associations, as well as by users and conformance testing organisations.
- Products and services conforming to the current specifications are available and have been demonstrated to work together.
- MAP and TOP are specifically aimed at providing communications standards which address the issues arising in implementing factory and office networks (Fig. 6.2).
- They are true multi-vendor standards that fully support interconnection of devices regardless of the manufacturer.

Some existing networking systems will be able to migrate easily to MAP or TOP

A new standard is fine for companies starting out in communications but it may be considered unfeasible for those which already have a heavy investment in proprietary networks. Fortunately, some vendors have for a few years been taking account of the direction in which the tide is flowing, and the new standards themselves reflect current practice to some degree, so the gap to be bridged may not be too great. The Ethernet specification developed jointly by Digital Equipment, Xerox and Intel, and adopted by many other companies, is very close to part of the MAP and TOP specifications, and the DECnet extension by Digital Equipment is almost the same as the TOP specification.

Most existing networks will be able to interface to MAP

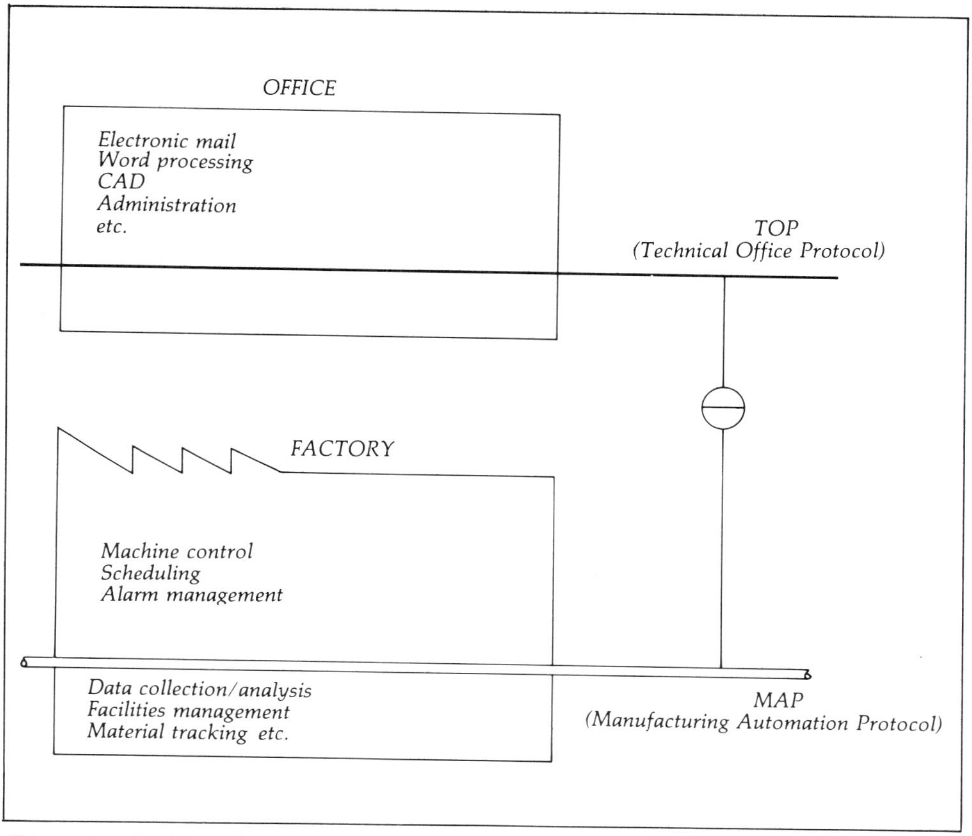

Fig. 6.2. MAP and TOP are complementary multi-vendor standards which are developing to serve the differing needs of office and shop-floor

and TOP via gateways – and indeed can use the MAP broadband cable as their physical link. Whether it is worthwhile to interface them will depend on a number of things, one of which is their closeness technically to the MAP/TOP specifications, which will determine the cost and operating speed of the interface.

New additions to MAP will offer lower-cost solutions for smaller companies

In its present Version 2.1 form, seen publicly most recently at the CIMAP demonstration at Birmingham in the UK (Fig. 6.3), MAP is filling a gap in the communications needs of large companies with a number of applications programs to be linked to each other and to

Fig. 6.3. The CIMAP demonstration at Birmingham in the UK in December 1986 showed that MAP is now a robust standard with a good range of products

automated manufacturing cells and systems on the factory floor. Smaller companies may see it as giving a more complex solution than their problems warrant, especially in view of the current price tag on a full MAP implementation.

Two things can be said about this. One is that prices will fall drastically as more users and vendors come into the MAP marketplace. The other is that forthcoming versions of MAP offer the prospect of simpler solutions which may be relevant to smaller applications. In particular, the carrierband version (see Appendix A) could prove attractive to smaller users as well as for lower-level applications in large organisations.

With so much publicity currently for MAP and TOP it is important to see the whole subject in perspective. A

MAP/TOP do not in themselves provide any ready made solutions to networking problems

company should assess its networking needs from the systems architecture which it has developed, as described in the last chapter. The systems architecture will show what types and volumes of traffic are required between systems in different parts of the company, and what the growth expectations are in the next few years. It is on the basis of this reliable information that decisions will be taken about the networks to be adopted for company communications. If the communications involve the linking-in of manufacturing equipment in the factory then MAP is of sufficient stature to merit serious consideration as the networking method, but one should not assume that MAP is automatically the best solution simply because it is the latest technical innovation.

7 BUILDING OPEN FACTORY NETWORKS

Because of its importance as the foundation of a computer integrated system for manufacturing, the design and installation of a comprehensive network system is likely to be an early step in the implementation of the systems architecture.

DESIGN

It is probably best to install the complete network at one time, allowing for future growth

The systems architecture and migration plan will have defined the relative positions of existing systems and networks, and those new services which are required. It is likely that a structure will have emerged which identifies both horizontal relationships between systems servicing different functional areas and vertical relationships representing coordinating levels within those functions (Fig. 7.1). The communications requirements of the various levels in the systems architecture will be different in each case, and the design of the factory network must take these differences into account.

It is most unlikely that the network system will be fully utilised until the systems architecture is completely implemented, but at least in the case of a MAP network it is probably best to install the complete network for a factory at one time, making provision for the additional nodes (connection points to the network) that will be brought into use at later stages as the information technology programme develops. Therefore design of the various networks in the system will probably be an early activity.

Network design is a complex specialist task which is best done by professional network designers. However, they will need to have a detailed specification of the company's requirements, and if the progress of the design is to be properly monitored and understood and the final design intelligently assessed, it is important to understand the considerations which enter into network design.

Network design must take many different factors into consideration

1. *Redundancy*. Even a brief failure in certain parts of your network might have serious consequences. It is important to think about this possibility and if necessary to ensure that the chance of such an event is reduced to a minimum by requiring some redundancy in the communications system – which is quite distinct from database redundancy.

In different circumstances redundancy may take the form of duplication of certain parts of a network, with parallel cable runs. It may involve duplication of key items of equipment like head-end remodulators in a MAP network.

Run the network to all areas where access is likely to be needed

2. *Take a long-term view*. Particularly in the case of a broadband MAP network, think carefully about the area it will ultimately cover and the number of nodes it will require. It is by no means impossible to extend or add nodes to a MAP network, but the cost is greater than with

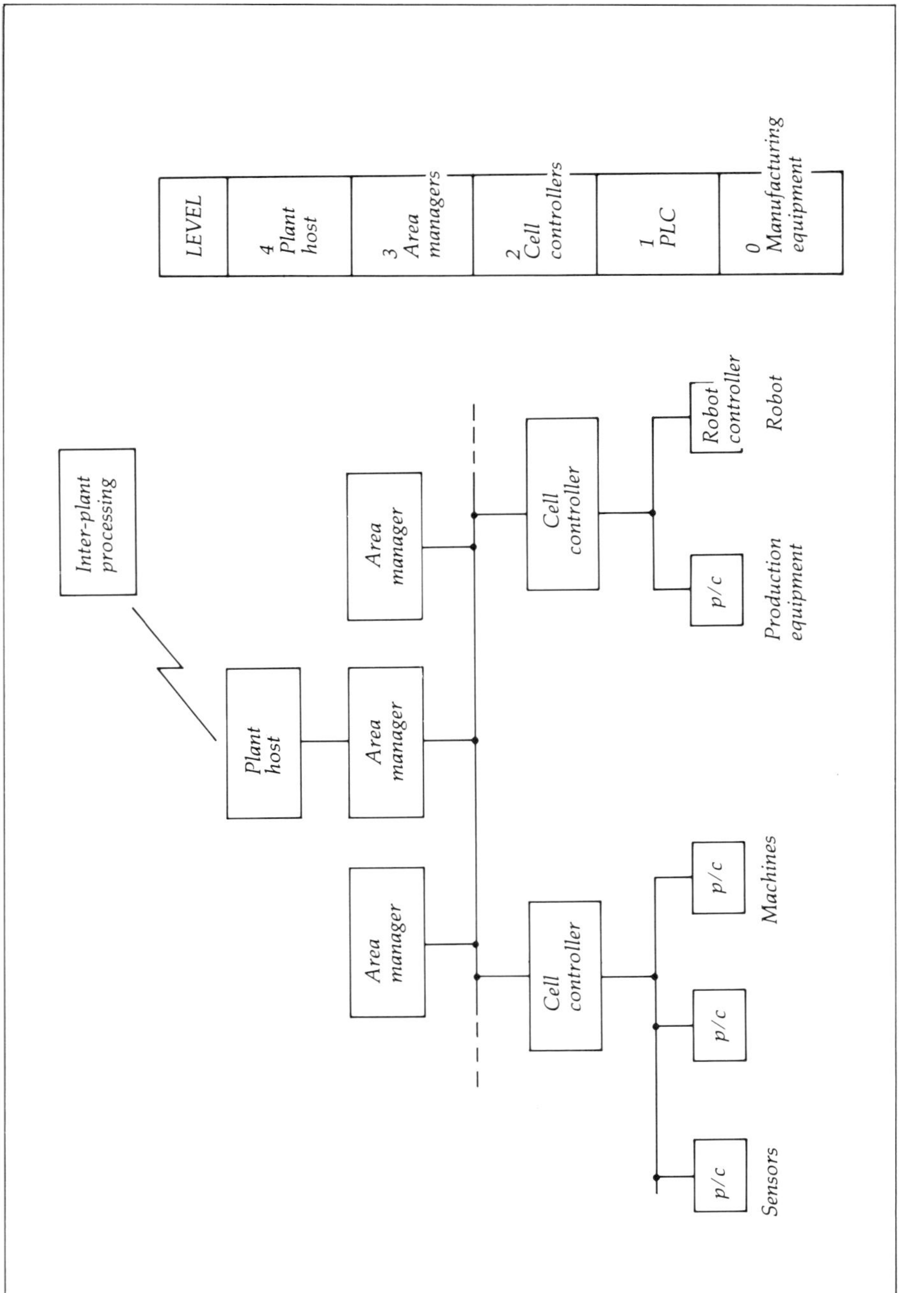

Fig. 7.1. The differing requirements within a company are often best satisfied by a hierarchical Systems Architecture

a TOP network because the network has to be retuned. The designers or installers will advise on the best procedure. But it is wise to run the network into all areas where access is likely to be needed, and to be generous with provision of attachment points for cable drops.

3. *Interconnection.* Whatever types of network are used in the overall architecture, make it a principle that every node should be able to connect to every other node. This may require the use of routers and gateways to bring proprietary networks into communication with MAP and TOP.

4. *Reliability.*

5. *Access to wide area networks.*

6. *Time response criticality.*

7. *Load* (number of nodes, frequency and volume of transmission).

One of the great advantages of broadband cable networks such as MAP is that they provide great flexibility to systems engineers during the migration from the existing architecture to the vision of the future

One of the great advantages of planning to use a MAP compatible network results from the multi-channel capabilities of the broadband cable upon which it is based. An important benefit from this multi-channel facility for the systems engineer is that it gives great flexibility when developing the migration plan. For instance it is possible to make point-to-point connections over the broadband using relatively cheap modems – indeed, many proprietary networks can be operated over broadband cable.

A special quality of broadband cable is that it can be used to carry many different services simultaneously (Fig. 7.2.) It can carry TOP, SNA and other networks. It can carry point-to-point RS-232 and other links. And it can also carry quite different types of services such as closed-circuit television monitoring circuits for surveillance. All these wider opportunities should be borne in mind when planning a broadband network.

Broadband cable can carry many services simultaneously – plan to use them

Broadband cable is also extremely useful when considering the problem of how to provide a single user-interface to many systems running on different computers. Just because two computers communicate over a MAP

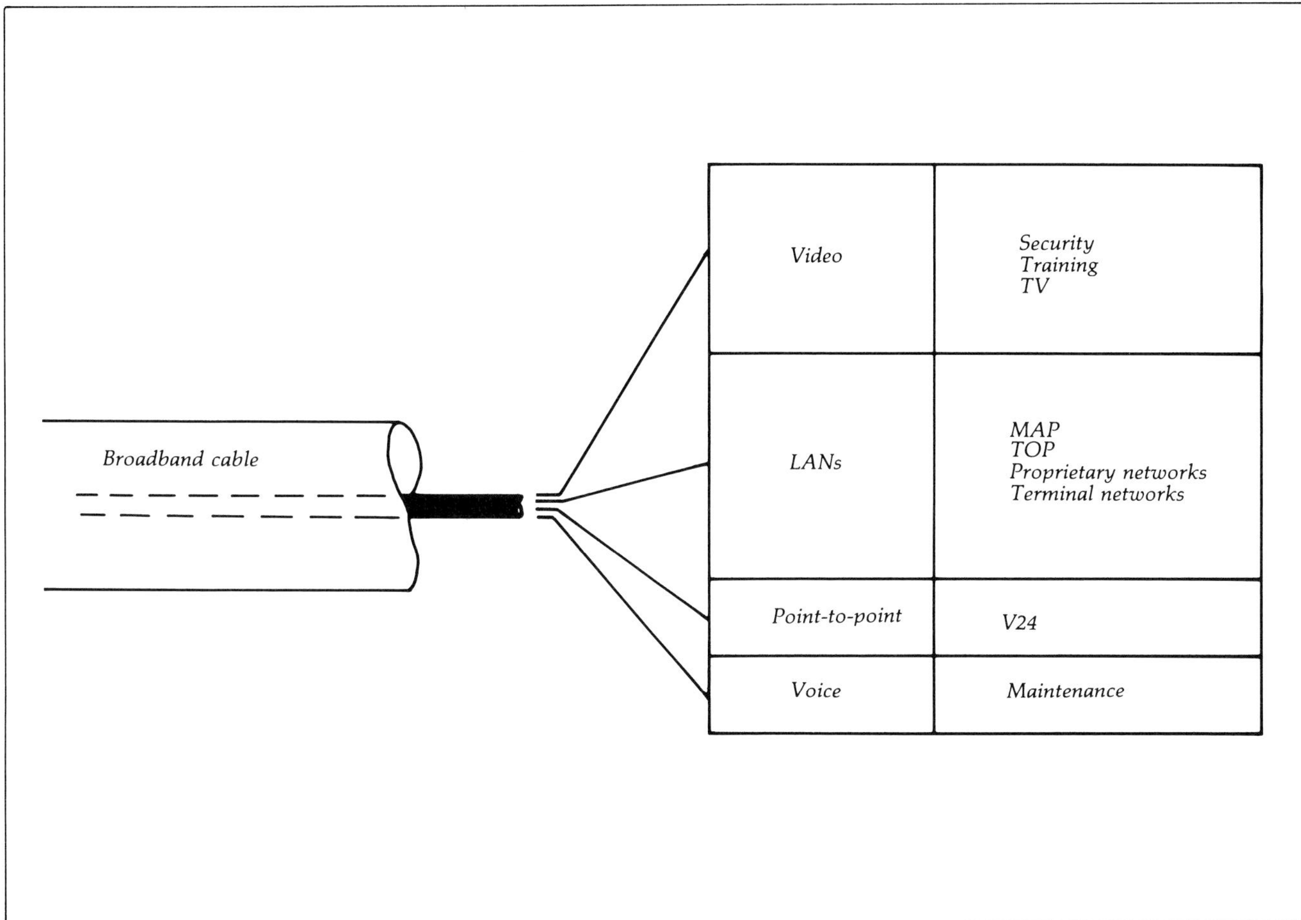

Fig. 7.2. Broadband cable can carry many types of services at the same time

network with each other does not mean to say that a terminal connected to one computer can access systems running on the second computer. Such limitations in the current capabilities of MAP present significant problems of designing a single user-interface. Broadband cable presents the systems engineer with the opportunity of running a 'terminal channel' which overcomes this limitation by allowing any terminal to be temporarily connected over the broadband cable with any computer.

Many companies are finding that the provision of a broadband cable network throughout the plant can often be justified on its own merits.

INSTALLATION

Three stages can be distinguished in the implementation of a network – and a MAP network is described here because it is the most complicated to install. TOP and proprietary networks are generally easier and less demanding. The three stages are:

- Network installation, tuning and testing.
- MAP system installation and testing.
- Applications installation and testing.

Employ professionals for network installation, tuning and testing

Network installation, tuning and testing are professional tasks, to be done either by the people who designed the network or by their subcontractors. Application testing, though, needs to involve the people who are going to run the applications, who will need to sign-off the system.

Tuning the network is a special requirement of MAP. After the cables and tapping points are installed an adjustment has to be made at each 'splitter box' where the cable branches and each 'tap' from which a spur is taken to a node to ensure that every node on the network will receive approximately the same signal strength. If the network is extended, or additional splitter boxes added for new nodes, the network will have to be retuned. This is not a very lengthy task, but the network needs to be out of service while it is done.

MAP itself requires the addition of certain facilities to the network. There must be at least one head-end remodulator, which receives outgoing signals from nodes in the network, checks them for errors, and re-transmits them at a different frequency for reception. There should preferably be a second head-end which is brought in by an automatic switch to take over if the first fails, and the operation of this needs to be tested during installation. Some MAP networks require the addition of network management software, and directory services software also needs to be carried in the network. Where the MAP network links with other networks, there will be routers or gateways to be installed and tested.

Add applications one at a time

With a MAP network the applications, with their interface hardware and software, are added to the network one at a time, starting with the installation and testing of two nodes. Each node has to be installed and its interworking verified before the next is added. There are four stages in the installation of each element in the network:

- Physical connection.
- Transport connection.
- Session connection.
- Application connection.

What is tested at these stages is the ability of each node to communicate with other nodes at the level under test. The four levels correspond to the principal communication layers defined in the MAP specification and corresponding to the layers of the International Standard OSI Reference Model. The transport layer connection verifies that packets of data can be sent successfully from one node to another and that the addressing system functions correctly. The session layer connection is a higher level test which may indicate that communication software may have to be modified to provide interoperability between nodes.

The final test is whether application programs at the different nodes can work together. This is the most difficult and lengthy stage, and every attempt should be made during these tests to discover any bugs in the software which could cause problems in operation. If an interfacing software system like Baseway is used, this has to be installed at an early stage.

Purchased equipment can be pre-certified for conformance to MAP

Testing of conformance to the MAP specification should not be necessary if equipment bought is certified as having been tested, but interworking of each application will have to be verified as it is installed. Tests are also needed of maintenance, security, and error recovery. It is very important to have a procedure for action if the network goes down. Particularly in the case of manufacturing cells there can be disastrous results if there is not a clearly defined procedure for restarting after a stoppage, which may require returning all programs and tools to a zero position.

8 INTEGRATING MANUFACTURING APPLICATIONS

Attainment of 'comprehension' is the most difficult and costly part of integrating applications

A networking system such as has been described in the last two chapters is essential in obtaining integrated communication, but it gives no more than communication without comprehension. If the software in one application area is to make use of information produced in another application, then both application programs must be persuaded to talk to the outside world of the network and to present the information needed in a form that is useful to the other end. With the introduction of standard protocols to simplify the communication task, this attainment of 'comprehension' is now the most difficult and costly part of the integration process.

If the applications are obtained from the same supplier

there can be some confidence that they will be compatible. They should define similar data similarly. If the software is recent then the supplier may have already provided for integrating applications by offering a consistent means for transporting data from one to the other.

Application software from different suppliers will generally require the building of special interfaces

If, however, application software is sourced from different suppliers, or is home-grown, then it is less likely to be compatible. In this situation there is only one practical way to make applications understand each other – by building interfaces between them (Fig. 8.1). An interface interprets the output of one application and creates the input for the other. The interface is independent of the applications themselves.

The main considerations in designing the interfaces between applications can be summarised under five headings:

- Functionality of the interface.
- Control of the link.
- Details of the data.
- Performance.
- Security and integrity.

FUNCTIONALITY OF THE INTERFACE

A network architecture diagram may show a line connecting MRP to Factory Control. That line conceals a highly complex relationship covering the release of orders on to the shop floor, giving a daily schedule to the factory controller and so on. In a manual system a great many user procedures and requirements are covered by those descriptions, and they do not disappear in an automated system.

Connecting a CAD system to MRP and production engineering similarly raises a great many questions. For example, how is the CAD system being used? What user procedures are associated with its use? Does, for example, the drawing office choose part numbers, or are part numbers given to it by some other function? If the latter,

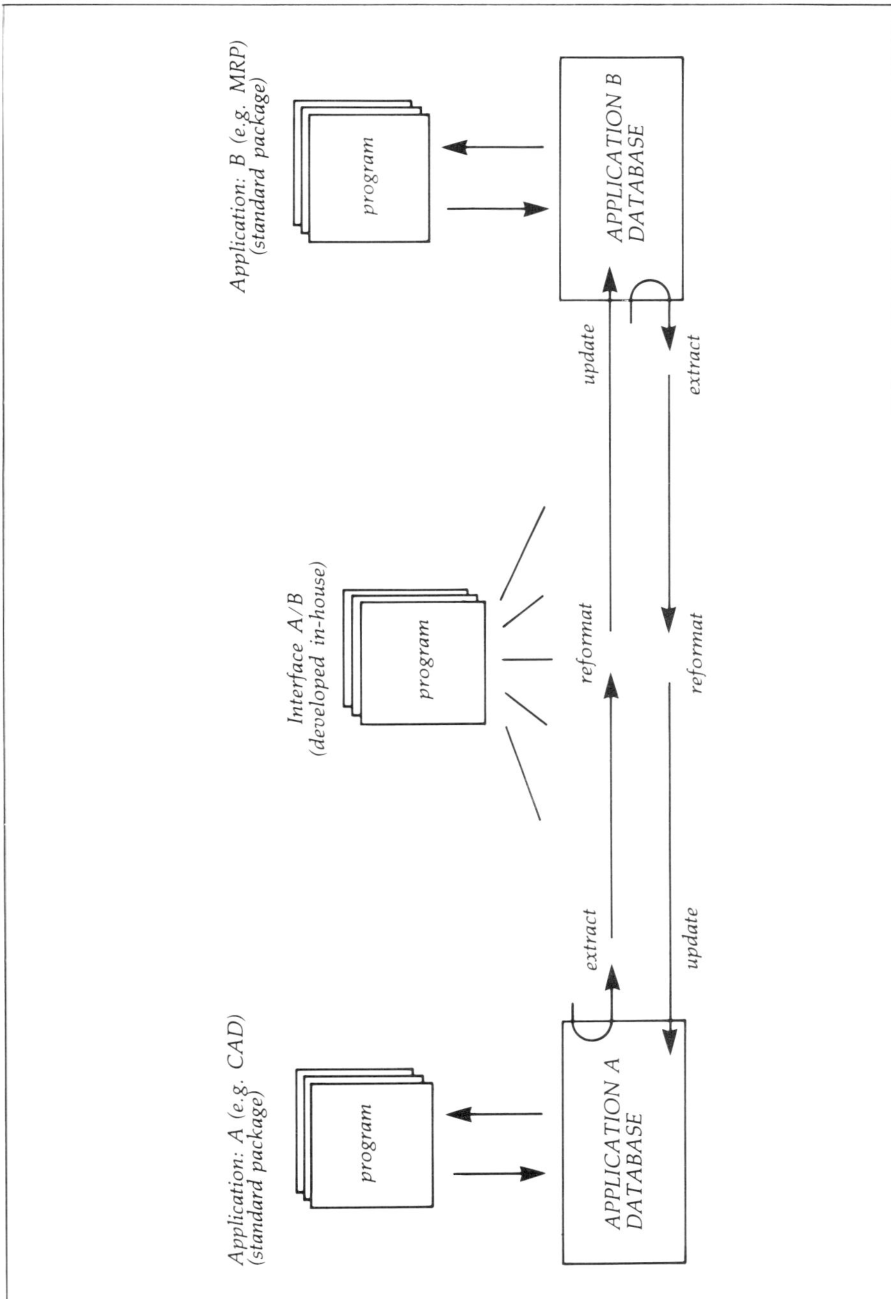

Fig. 8.1. Interfaces are developed to pass data between imcompatible systems. These interfaces should be as independent of the application programs as possible

what is that other function? Does it have a system? Some companies, for example, use a part numbering method based on a coding system indicating function, geometry, material and other characteristics. Systems analysis has to be done on what the requirements are and how they are being met by the particular application.

If the operation of the application software is not accessible, it will be difficult to integrate

Look at the application itself. Is it at all possible to integrate it? To keep to the CAD example, is there a facility for extracting from the CAD output data in a form which can be put into a production engineering system such as tool design or NC part programming? If the information is not accessible the application is going to be difficult to integrate. On the design to production control link, can the CAD output produce a bill of materials? Many systems cannot. If parts lists are simply given as graphics information on the drawing there is no simple way in which it can be extracted.

At the other end of the link, are there input mechanisms? Does the bill of material processor accept alien input or does it expect the data to be typed in at a terminal keyboard? If so a background program will have to be written, capable of updating the database with information from the design system.

Consider the functionality of the link itself. What user function does it represent? In the design to production interface, for example, there are quite often very complicated user procedures covering the release of designs to production. A level of confidence may have to be assigned to each component before it is released into production. The drawing may have to be signed off by several different authorities. Then it may be necessary to check out the suppliers and to verify that those specified are on the approved list.

Decide how much automation is appropriate in triggering the transfer of information

Another batch of questions relating to the functionality of the link concerns the triggering of information transfer. In the case of MRP orders to the factory controller it will be the day's orders. What happens in the existing system to make us give the orders to the factory controller? What about priority orders? What degree of automation are we prepared to see in the working of this interface? How can the software be built flexibly so that if one of the systems is changed the interface will still work? If the CAD system is changed, for example, it may alter completely the way the parts list information is extracted from drawings.

Decisions must be taken about reporting of information flow

What kind of reporting is needed? Is it important to know whether a transfer has worked or not? The reporting has to be relevant to the information flow. If it is an on-line link there need to be on-screen messages. If it is to be a highly integrated system, how do we manage the flow of information into all the different interfaces? If, for example, there is to be a direct link between CAD and Production Control, and at the same time a link via Production Engineering, so that one way the design is picking up materials information and the other way routing or methods are being established, how do we make sure that both the material and the routing information are ready at the same time so they can be used by MRP and given out as orders the next day to the factory controller?

None of these are insoluble problems, but it is important that they should be addressed.

CONTROL OF THE LINK

Under this heading come questions of 'ownership' of the link between applications. In a CAD to Production Engineering link, for example, one has to ask questions like:

- Is it the draughtsman who sends the drawings or the production engineer who asks for them?
- Which system initiates or controls the transfer of data?
- In which direction is it going – in both directions? How often, and how much information is being transferred?
- What kind of network links therefore need to exist? Can they be handled simply by the two operating systems or must the application programs be concerned?

Types of file transfers required will have a bearing on the choice of network facilities

Transfer of files is normally from one machine's filestore to another machine's filestore, which is controlled by the operating system. If messages are being transferred in real time between two application programs, it is the application programs which have to deal with it as well as the operating system. If it is program to program transfer, there are questions of how the programs use or gain access to the top layer of MAP or TOP, and what programming needs to be done to make use of MAP's manufacturing messaging facility. Study of these questions will give some idea of which network facilities it will be most appropriate to use.

DETAILS OF THE DATA

If an application program is unable to provide information needed by another it probably has a serious software inadequacy

The questions here concern the compatibility of data formats and data structures in cooperating applications. How much information is involved? How is it held? What is its format? How can we translate the data from one format to another? What is its structure? If the link is to be between CAD and downstream functions, are the part numbers used in CAD also used in Production Engineering or Production Control? If they are different something will have to be done about it. Are we talking about transferring single-level parts lists or multi-level bills of material? Is there a difference between the type of list needed by Design and the type needed by MRP? They could lead to radically different application programs. Is there information needed by one application which it is impossible for the other to provide? If so, there would appear to be a serious inadequacy in the software.

PERFORMANCE

Important things to look at under this heading are the response time required and the volume of data to be handled. Overnight file transfer may be adequate for a CAD/MRP link but it would be useless in real-time shop-floor control. If very fast response is required it may be necessary to look at the new extensions to the MAP specification such as Enhanced Performance Architecture and Mini-MAP. The frequency of access and transfer is also relevant here.

When calculating the volume of data to be handled it is important to make allowance for growth in the future.

SECURITY

The more technical problems relating to security concern integrity of data. A classic problem is that of distributed databases – where you have, say, a design system and a production engineering system and are transferring a single level parts list from design to production engineering, and the network breaks down half way through the transfer. How do you ensure that when it starts up again you don't have half the data at each end with neither system aware of the fact?

It is essential to ensure that information is not lost between applications during updates

Problems like this, where an update is dependent on two nodes in the network are in a category called 'commitment control'. System A is saying 'I want to update System B and I want to know that I have updated it'. System A starts the update, sends the information to System B which starts the update. System B cannot 'commit' until System A has completed the transfer and signalled that it is complete. Until System B commits, both systems can revert to the state before the transfer was started, which is necessary to preserve data integrity.

Restarting of equipment after emergency shutdown presents special problems

Error handling is another aspect of security. What happens when things go wrong? How fast must they be reported? Does the user always need to be told? At the application level, what are the procedures for correcting errors? Particularly with automated production equipment like robot lines and flexible manufacturing systems, restarting after a system fault can be a very complex task. If the system is integrated with management reporting, quality control and maintenance there are other aspects of error reporting to be taken into account.

If you have gained the impression that integration of applications software is a highly complex task you will be right. It is not impossibly difficult now that the lower-level networking problems have been dealt with, but except for a very large company with a lot of in-house expertise it is a task best handed over to specialist systems integrators.

If source code of an application package is not available it may be best discarded

There are situations where it may be better to discard an existing application package rather than attempt to integrate it into the system. If it is necessary to modify the application software in order to integrate it various difficulties may arise. The source code for the software may be lost, its origin may be unknown, or its suppliers may be unwilling to make it available. In any case the amount of rewriting required may not justify the expense of what may ultimately still not be an entirely satisfactory product. It is commonly estimated that systems integration costs about 60% of the total for an integrated system as against 40% for the networking costs, and this is a quite typical ratio. Integrating a system between two application programs can be a bigger and much more difficult task than carrying out a complete MAP implementation.

There are signs that the work of systems integration will itself become easier during the next few years. We can expect to see more progress towards standard ways of defining and formatting data along the lines of the manufacturing messaging service which is already incorporated in the MAP specification. There is room for more standards covering data definition, of which the International Graphics Exchange Specification (IGES) is an example. Another under development is the Trade Data Exchange standard, which is concerned with describing and transmitting the information which goes into orders, invoices, price lists, goods received notes and similar paperwork. There is also discussion of a product definition description standard which would cover the way a parts list is presented.

Software packages are under development which will allow easier interfacing via MAP and TOP

Some companies are also working towards the formulation of software packages to provide a 'black box' interface between the top of MAP or TOP and individual application programs, which could be installed by nonspecialist personnel without too much difficulty. Now that the networking difficulties are virtually out of the way we can expect more and more people to turn their attention to the subject of systems integration.

9 COSTS AND COST JUSTIFICATION

The totally integrated manufacturing organisation is a beautiful vision, but real companies in a real world have to see a return on their expenditure, and in the UK if not elsewhere they have to see it very quickly. This chapter will look first at the costs a company will incur in installing a network and integrating computer applications on it. Then it will review the benefits to be expected from the integration of manufacturing applications. Finally it will look at the process of carrying out a cost justification exercise for an integration project.

The overall cost of an implementation depends on the complexity of the network and the amount of in-house or contracted system integration expertise required. However, it is possible to estimate the price of particular devices and to indicate a 'typical' breakdown of costs that might be expected, under a list of headings like the following:

- Computers.
- Communications technology (head-ends, cable, network managers).
- Terminals.
- Application software, enhancements and interfaces.
- Support and maintenance.
- User time.
- Systems integration development.
- Operational software.

Investments in these areas of technology must meet the same cost justification criteria as other types of investment projects. There are, however, some special features in these advanced manufacturing technology projects which must be taken into account. Ignoring the 'intangible' benefits which, though real, are difficult to quantify, if any investment in manufacturing is to reduce costs it must have an effect on at least one of these cost factors:

- Material – either its type or its use.
- Yield or utilisation.
- Energy.
- Control or cash.
- People or indirect overheads.
- Facilities overheads.

In the past, specific investments tended to be targeted directly at one or at most two of these factors of cost – usually direct labour. CIM tends to affect many if not all of the factors of cost, and therefore its impact is far more difficult to evaluate.

At first sight, for example, one might see the installation of a factory local area network as perhaps saving the time of two or three people progress chasing materials or NC programs across the factory, saving perhaps £72,000 in three years against a network cost of £60,000. But then a technical specialist will be needed to look after the networking, and the two people displaced will probably be needed for something else, so that considering only these immediate consequences the network does not appear to justify itself.

But the introduction of a network will affect other factors of cost. About 30% of product cost on average goes into ensuring its quality. If a networking system improves the quality of data and gives greater consistency in products because the data does not have to be recreated in transfer from one area to another, then some very big areas of cost are being hit. These are not 'intangible' values, because the savings will be measurable in indirect overheads and probably in better material utilisation – two of the factors of cost already listed. It will require skill and experience to put a reliable figure on these savings, but the consequences will be real enough.

With careful planning it should be possible, even with an initial pilot project, which may have to carry the full burden of the cost of installing the network, to show a positive return, though the project should be assessed as part of the company's longer-term information strategy which will include further investment projects.

Integrated manufacturing projects should not, therefore, be considered in isolation from the other projects proposed in the systems architecture, even though each project should be individually cost justifiable. The systems architecture document described in Chapter 5 will have provided a broad plan for implementing the company's information strategy over the next few years, with an outline of the steps by which it should be achieved. Chapters 6, 7 and 8 have reviewed some of the more detailed technical considerations that must enter into the business planning of the implementation programme. In parallel with the technical work it is also necessary to carry out the financial assessment, and Fig. 9.1 shows how these two streams flow from the project implementation plan.

The first stream is the financial model itself. There are well-established techniques for appraising capital investment, but not all of them are equally appropriate to project which form part of a longer-term programme. It is essential in this type of justification to use a total company

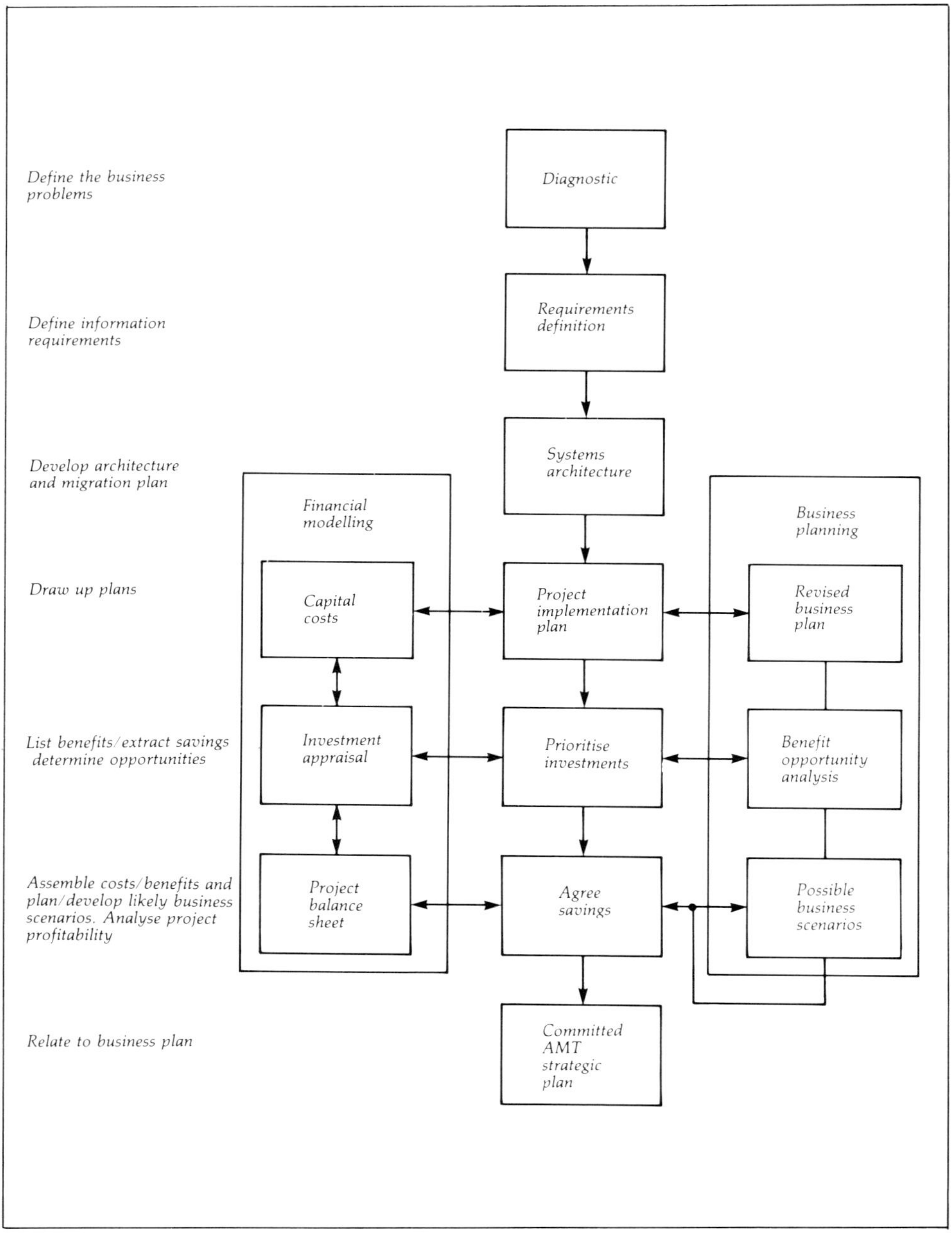

Fig. 9.1. Business plans should interact with financial considerations in deciding on CIM projects

model, and given the long time phasing of cash outflows and inflows, the most pragmatic approach appears to be internal rate of return, with risk-adjusted internal hurdle rates. Certainly payback methods do not provide a realistic basis for evaluation of these projects.

Alongside this mechanistic financial modelling the business plan spells out the ways in which the projected investments will be followed through to produce the expected benefits.

The *project implementation plan* takes the migration plan represented in the systems architecture document as its starting point, and develops the full details of the stages, tasks and actions which together will comprise the total programme. It will contain careful estimates of the man-days and skill levels required to develop and implement the project. This plan will be constantly modified during the investment appraisal with respect both to its timing and to its resourcing, as operational departments identify the constraints upon their operation and on the individuals who will staff the project. It is helpful to express this plan on a simple computer-based project planning system.

Prioritising the investments is based on the diagnostic work previously completed and the opportunities identified during requirements definition. The key driver of investment priority, however, is usually the outcome of the investment appraisal model. This will show how to carry out the projects - as far as technically feasible - in a sequence which will maximise the early cash inflows.

Agreeing and committing to savings is potentially the most difficult activity in the justification process. Agreeing and committing to savings by line management requires their awareness of the full implications of the investment programme, and hence requires education. It is useful to view the budgets of operational departments in terms of their major factors of cost, as listed above, and to balance these costs against benefits resulting from the programme as seen from the viewpoint of the business as a whole. Frequently it is impossible to reconcile these two perspectives, in which case it is useful to define both a committed and a potential target, and to carry out the financial analysis on both bases.

This method of appraisal and cost justification takes into account the costs and benefits to be derived from the whole of what may be a five-year plan containing a number of individual projects. Each project will have been assessed individually, but within the framework of the whole plan. It will be wise to study in most detail the projects which are expected to be implemented first, because a re-assessment may well become necessary during the course of the programme, either as a result of changes in the industrial marketplace or in the light of experience with the initial projects.

The justification of computer integrated manufacturing projects is complex, but it is not impossible. It requires a carefully structured approach such as that described here, and a combination of both a top-down and a bottom-up perspective which examines the functions of both the technical and the business justification – which may well be different. It ultimately requires both awareness and education. For this reason it is difficult to plan how long it will take. But in the experience of Coopers & Lybrand the results of a careful justification process are frequently positive, and if properly undertaken provide the foundation for a successful project.

APPENDIX A –

INTRODUCTION TO MAP AND TOP

Before application software such as a material requirements planning system can send scheduling data to an automated cell in the factory many conditions must be satisfied, among them:

- There must be a communication link capable of conveying messages and commands between the two systems.
- There must be facilities to ensure that every message gets to the correct destination, that it is complete and without errors.
- There must be the ability to send information files between systems which may structure their files differently.

- There must be a common format in which, for example, commands to a robot controller are expressed.
- It must be possible to extract from the mass of data held in an MRP system just the information necessary for a cell controller to schedule its work.
- There must be the ability for the cell controller to report back significant variations from the schedule to update the MRP system.

Only the first four of these requirements are covered by the MAP and TOP specifications. The others - and many more besides - have to be handled by the communications software which sits between the individual applications and the network which connects them. Before MAP and TOP, achieving the first of these was a major, if not insoluble, problem.

MAP and TOP are necessary elements in a cost-effective computer integrated manufacturing system, but they represent only a minor part of the cost and effort involved in setting up the links.

HOW MAP BEGAN

The Manufacturing Automation Protocol began life in 1980 when General Motors in the USA set up a task force to identify communications standards which would allow data communications between equipment and software from different vendors. You can understand the reason for this initiative from the fact that in 1984 the company had some 20,000 programmable logic controllers and 2,000 robots. The number of intelligent devices in the manufacturing area totalled more than 40,000 with a projected 400-500% increase in the following five years. During that period GM expected to spend hundreds of millions of dollars on applied computer technology.

According to Michael A. Kaminski Jr, manager of MAP at GM, only 15% of the 40,000 programmable tools, instruments, controls and systems installed in GM are able to communicate with each other, and when such communication does occur it is costly, accounting for up to 50% of the total expense of automation, because of the special wiring and the custom hardware and software needed.

THE OSI MODEL

The task force quickly recognised that its best course of action would be to recognise and promote an international set of standards which were already being put in place within a framework known as Open Systems Interconnection (OSI). The OSI framework is called a 'reference model', and was developed under the sponsorship of the International Organisation for Standardisation (ISO). It defines the way in which the essential elements in a total communications system should be distinguished and how they should connect together.

The OSI model had to be comprehensive enough to cover all types of communications, from local area networks in offices to international public data transmission services, and including the needs of industry, commerce, the media, governments, and so on, for transmitting text, numerical data, graphics, photographs, digitised speech, and other data. Within the OSI framework many different standards have been and are being established reflecting these diverse needs, but because they conform to the reference model the task of creating links between them is made much easier.

What the GM task force did was to select from the standards conforming to OSI a set which were suitable for the task of factory communications. When the task force began its work, most of the standards were still only in draft form, and for some of the requirements the standards had not even been drafted, so part of its work was to help in expediting the establishment of standards. Even when the current Version 2.1 of the MAP specification was first published it had in many cases to refer to draft standards or to specifications formulated by other organisations such as the US Institute of Electrical and Electronics Engineers (IEEE) which were expected to be adopted as International Standards. Only in limited areas, however, did the task force have to fill the gaps by drawing up its own detailed specifications.

Today the MAP specification is backed by a large number of user companies around the world, and most of the leading equipment and software suppliers in the manufacturing arena have declared their intention of supporting it. Its progress – because the specification is still developing – is helped along and promoted by user groups in the USA, Europe and other parts of the world.

HOW TOP BEGAN

The Technical and Office Protocols specification (TOP) has a shorter history though similar origins. In this case it was the aerospace company Boeing which initiated the movement and is spearheading its development. From the first, TOP has intentionally been closely related to MAP – and hence conforms also to the OSI reference model – and the two specifications are now developing jointly. TOP is aimed at a rather different range of applications, but there is some overlap and the two types of networks can easily interface with each other.

TOP is primarily intended to serve the communications needs of offices in manufacturing companies – electronic mail, word processing, editable text and non-text document exchange, file transfer, graphics interchange, database management, distributed batch jobs, videotext and business analysis. Apart from its interface to the outside world and in the lowest levels of its communication functions it is identical to MAP.

The details of the MAP and TOP specifications are not of great importance to the network user except insofar as they require the installation of certain types of cable and particular types of equipment, and offer certain interface facilities to the outside world.

The OSI reference model – and hence the MAP and TOP

specifications – has seven layers, which were illustrated simply in Fig. 6.1. The International Standards adopted by MAP and TOP are summarised in Fig. A.1. All except the top and bottom layers are tucked away in the software or firmware of the MAP or TOP interface, and only the

ISO reference model with MAP/TOP common core of protocols		
LAYERS	*TOP V1.0 PROTOCOLS*	*MAP V2.1 PROTOCOLS*
Layer 7	*ISO FTAM (DP) 8571 File Transfer Protocol*	*ISO FTAM (DP) 8571 File Transfer Protocol, Manufacturing Message Format Standard (MMFS) and Common Application Service Elements (CASE)*
Layer 6 Presentation	*NULL (ASCII and Binary Encoding)*	
Layer 5 Session	*ISO Session (IS) 8327 Basic Combined Subset and Session Kernel, Full Duplex*	
Layer 4 Transport	*ISO Transport (IS) 8073 Class 4*	
Layer 3 Network	*ISO Internet (DIS) 8473 Connectionless and for X.25 — Subnetwork Dependent Convergence Protocol (SNDCP)*	
Layer 2	*ISO Logical Link Control (DIS) 8802/2 IEEE 802.2 Type 1, Class 1*	
Layer 1	*ISO CSMA/CD (DIS) 8802/3 (IEEE 802.3) CSMA/CD Media Access Control 10Base5*	*ISO Token Passing Bus (DIS) 8802/4 (IEEE 802.4) Token Passing Bus Media Access Control*

Fig. A.1. How MAP and TOP conform to the ISO seven-layer model

access to layer 7 is encountered by the user. This is where MAP and TOP will differ in the services they provide.

CABLES FOR MAP AND TOP

At the bottom of the stack at layer 1 are the only other features which differ in MAP and TOP. MAP Version 2.1 specifies the use of a broadband coaxial cable to provide the physical communication between nodes. TOP Version 1.0 is less dogmatic. It can be run on a broadband cable, but it can also use the lighter-weight coaxial cable which is used in Ethernet networks and often referred to as baseband cable. Both expressions 'broadband' and 'baseband' are used here rather loosely because strictly they describe the method of transmission rather than the cable carrying the transmission.

Broadband cable itself is fairly rigid. In its usual 75ohm variety it is ½in. diameter and requires a minimum bend radius of 6in., which must be taken into account in planning cable runs. Amplifiers can be used in the cable to augment the signal over longer distances, but where very long runs are needed between amplifiers a larger ¾in. cable can be used. In a factory the cable will probably be run in the roof, though in offices it may be preferable to run it under the floor. It incorporates its own shielding and is relatively safe from electromagnetic radiation fields. Additional shielding is only necessary where it comes close to production machinery. The broadband cable can also carry its own power supply for devices on the network like head-ends and amplifiers (Fig A.2).

At each place where equipment is to be connected to the cable a device called a tap has to be inserted into the cable. This is like a T-piece which carries one or more spurs of cable to the nodes to be connected. The cable has to be cut to insert each tap, and when complete the network has to be tuned. This involves making adjustments at each tap to obtain the same signal strength at each node. It is these considerations which make it desirable to design and install a complete MAP network as a single operation, though it is

Fig. A.2. Broadband cable and amplifier connected to head-end equipment

by no means impossible to extend or add to an existing network.

MAP nodes send out messages at one frequency and receive them at another. The conversion from one frequency to the other is carried out by a device called a head-end remodulator, and there has to be at least one in each MAP network. If reliability is a vital consideration it may be desirable to have a second remodulator in the network.

Also somewhere in the network there needs to be a computer which can carry out network management functions using special software covered by the MAP specification. A dedicated microcomputer can be used for this purpose, or network management can be carried as one of the functions of a minicomputer or mainframe connected to the network.

A TOP network is rather simpler to design and install. It

can run on a broadband cable, but a lighter weight flexible coaxial cable is all that is necessary. Taps for this baseband cable can simply be clamped in place after a hole is drilled in the outer insulation, and a connector is forced into contact with the centre conductor (Fig. A.3). Connected to the tap is a device called a transceiver which deals with transmission and receiving of signals through the network and carries out the carrier sensing and collision detection functions required by this type of network, described below.

Broadband technology is well established and has been used for many years outside manufacturing industry to carry a variety of services in airports, hospitals and the oil industry. For instance at Heathrow Airport, London, British Airways has a broadband network extending over 10km throughout workshops, test areas, stores and services, as the basis for a real-time information system designed to deal with all aspects of aircraft maintenance

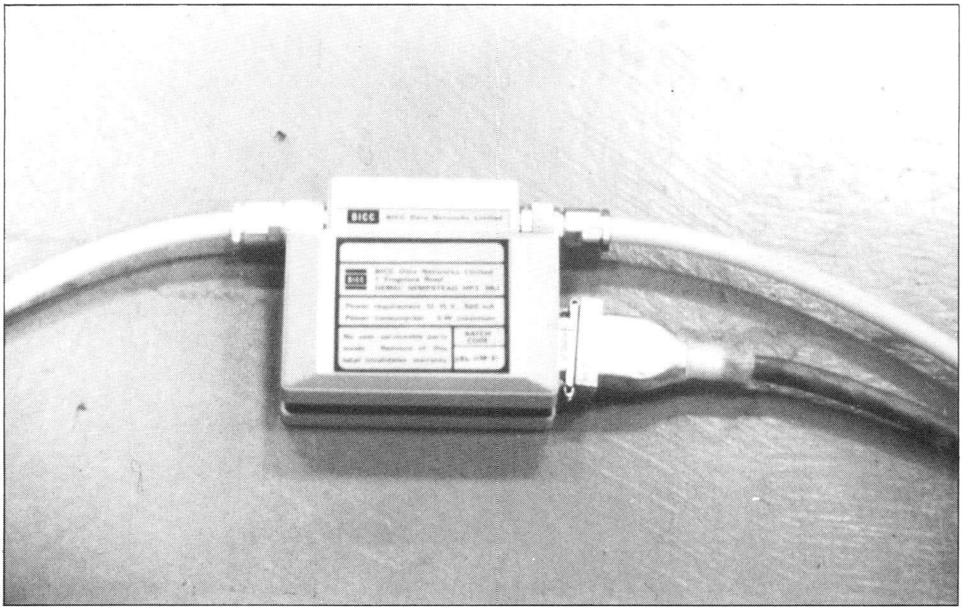

Fig. A.3. Transceiver unit for TOP network showing baseband cable (top) and drop cable (bottom)

and ground services. It connects a central IBM 370 mainframe with over 70 Ferranti controllers and more than 500 connected visual display units, keyboards and printers.

As a result of experience with major installations such as this the cable, connectors and basic networking equipment are well tried and tested, giving the user confidence in the reliability of this level of the MAP system.

GAINING ACCESS TO THE NETWORK

The clear distinction between MAP and TOP at the lowest levels is in the way different nodes such as CAD terminals, machining cell controllers and mainframe computers gain access to the network. The problem is that only one pair of nodes in a single network can communicate at one time. If another node started sending messages at the same time the result would be chaotic. But instead of letting one pair of nodes occupy the network for what may be a quite lengthy conversation, the message is broken up into small packets which are sent one at a time, giving other nodes the chance to get a word in. The task of seeing that every packet reaches its destination correctly and in the right sequence is the responsibility of layers 3 and 4 in the OSI stack.

MAP uses a system called 'token passing' to give nodes access to the network. This treats all nodes fairly, giving each in turn the right to make use of the network by passing round a 'token', which is simply a brief electronic signal. If a node does not want to communicate it passes the token on to the next node on the list, and the last node on the list returns the token to the first. If a node does want to communicate, it 'holds' the token and is allowed to transmit for a limited period of time. The method ensures that everybody has a turn and that each node can be sure of getting the token within a fixed period of time. It was this deterministic feature which GM considered to be essential for communications which may be controlling the operation of production machinery.

TOP on the other hand gives every node a sporting chance of gaining immediate access to the network whenever it asks. A node listens to find if the network is busy. If it is not, the node transmits it message. It is possible, though, that a second node will also detect that the channel is free before the first node actually starts transmitting. If this happens both nodes detect that a collision has taken place, and both back off and try again after a random time interval. This method, which is given the jawbreaking name of Carrier Sense Multiple Access with Collision Detect (CSMA/CD) makes for faster communications as long as the demand for access is limited, but as the amount of traffic increases there are more and more collisions and the efficiency of the system begins to fall off rapidly. The effect of load on access time for the two systems is illustrated in Fig. A.4.

There are other differences which affect the cost of networking. The broadband system called for in the MAP Version 2.1 specification allows many different signals to be carried simultaneously at different frequencies. A MAP network only uses two frequencies - one for transmission and the other for reception - so other frequencies are

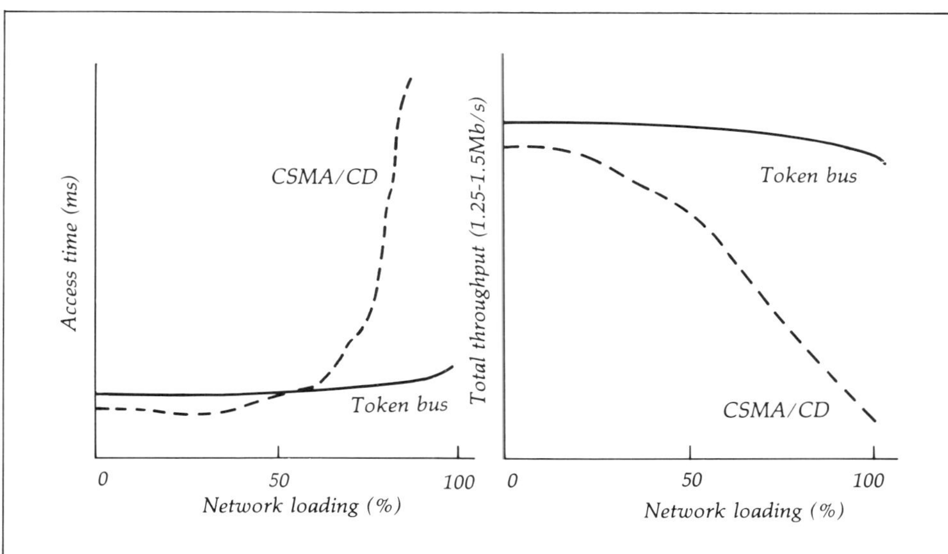

Fig. A.4. How access time varies with load using 'probabilistic' CSMA/CD and 'deterministic' token passing

available for different purposes, which could be other networks, including MAP networks, and point-to-point communications of various kinds. However, there is a price to be paid for this multi-channel capability. The transmissions along the cable are in the radio frequency range, which requires special equipment for decoding at each end which is rather more expensive than that needed in systems where the cable carries only one message at a time. Also MAP's use of two frequencies demands the use of what is called a 'head-end remodulator' at the end of the cable to receive the transmitted messages and convert and retransmit them at another frequency for reception. For added security a network may have two such remodulators.

TOP and other networks can also make use of the broadband cable with a suitable interface, and do not themselves require a remodulator since they only operate at one frequency. But TOP can alternatively employ a lighter-weight baseband cable which carries only one signal at a time at less than radio frequencies.

Complicating things a little further, in the latest MAP specification, Version 2.2, is provision for a carrierband version. This is similar to the broadband specification, including the use of token passing, but it does not require a remodulator. If desired, it is also possible to use a thinner baseband cable, though one loses the capability of using the cable for multiple services.

Carrierband MAP offers lower cost than the earlier specification, with better noise immunity than baseband systems. It can also be used with a further extension in the Version 2.2 specification known as the Enhanced Performance Architecture, which allows faster data rates.

APPLICATION INTERFACES

At the top of the OSI stack is layer 7, which interfaces with a company's applications like CAD, MRP and production machinery, via any intermediate software which may be required. MAP and TOP share with all other OSI compatible systems a package of services known as the Common Application Service Element. This contains facilities which will be required in any communications activity.

Also available within the OSI framework will be a number of Special Application Service Elements for particular types of activity. Some of these are already defined: others are still evolving. MAP and TOP share one of these elements, known as File Transfer, Access and Management (FTAM). This is designed to allow files to be transferred between different types of systems, and also to allow some degree of manipulation of files and parts of files. Different computer and software systems structure files in quite different ways, making it impossible for one system to read files from another. What FTAM does is to interpose what is called a virtual filestore which provides as it were a common language into which the filing systems at each end can be translated.

MANUFACTURING MESSAGING

A special requirement of MAP, which must communicate with shop-floor controllers of flexible machining systems, robot paint spraying lines and other automated equipment is to have a common terminology for messages and instructions appropriate to these types of equipment, which can be understood by cell controllers from different manufacturers. There was, and still is, no International Standard covering this specialist need, so General Motors took the initiative, in collaboration with a committee of the Electronics Industry Association, in devising a Manufacturing Message Format Standard (MMFS) which was published at the end of 1984 and is incorporated in the MAP Version 2.1 specification. It defines a standard format for messages and a standard set of meanings for the components of the message. MMFS is now being offered by some vendors as part of their MAP packages, and was demonstrated in action at the CIMAP event in Birmingham in December 1986. An example of a MMFS message is shown in Fig. A.5.

Message:

GC:04	*20TN:*	*CMD*	*WDL*	*DS:8192*	*...data...*

Equivalent Hex codes:

02 04	*IF*	*08*	*09*	*05 20 00*	*.........*

Explanation:

GC:04	=	*Group count grouper indicates four more fields in message*
TN:02	=	*Transaction number 02. Used to correctly associate responses with requests*
CMD	=	*Command not requiring a response*
WDL	=	*Download write. Note: multiple download writes are allowed, when the maximum message length is not great enough for the entire download*
DS:8192	=	*Data stream follows, with the size of data stream equal to 8192 octets*

Fig. A.5. A MMFS message sending a download-write transaction from a computer station to a PLC carrying the entire program to be downloaded in the data stream

However, it was acknowledged from the outset that MMFS was only a provisional stopgap standard, and work has been continuing in the EIA committee toward the finalising of a new MMS (Manufacturing Messaging Service) standard as specification RS-511, which is expected to be adopted as an International Standard. This new standard will be adopted in the new generation Version 3.0 MAP specification due to be published in 1987. It is expected to differ substantially from the present MMFS standard.

Manufacturing messaging does not need to form part of the TOP specification, but there are other services which will be important in an office environment. At present only FTAM is implemented in TOP, but it is intended in future to add facilities for message handling, document revision and exchange, graphics, and database management. Facilities for network management and directory services, outlined below, have not yet been defined for TOP, though proprietary software can perform these functions as it did in the CIMAP event.

NETWORK MANAGEMENT

Network management is not a human job description but a set of software tools which can be used to monitor traffic on the network and report on any troubles like overloads or crashes, to admit new nodes to the network, and so on. More precisely it provides configuration management, performance management, event processing, and fault management. This is provided for in the MAP Version 2.1 specification, as also is a read-only version of 'directory services'. The aim here is to provide the network equivalent of a telephone directory of all the software programs accessible to users of the network. When one application program needs to refer to another, it 'looks up' its address in the directory. This facility makes it possible to move software, databases, and so on, physically around the plant and connect them up to different nodes without having to search for all the places where other programs refer to them in order to change the addresses. A revised directory entry for each change is all that is needed.

MIGRATION POLICY

'Migration' is an important word in the MAP vocabulary. It describes the process by which users of today's networking equipment will be able to keep in touch with future developments and not find themselves with a lot of obsolete kit – a very important consideration in making the new standards acceptable. There are two aspects to this process. One is the need to be able to connect existing proprietary networks to any new MAP or TOP network. The ease with which this can be done depends on how closely the network currently in use conforms to the OSI reference model. Some sort of 'gateway' will be required to make the conversion between the protocols, and if they are widely different this can be expensive and slow. Some vendors have announced their intention of migrating their products into conformance with MAP, without losing touch with the networks currently used by their customers.

The other aspect of migration is the continuing development of the MAP specification itself. In an area where technology is moving so rapidly it would be unwise to attempt to freeze the MAP specification for too long a period. On the other hand some degree of discipline is necessary. When General Motors published the Version 2.0 specification in January 1985 it undertook not to introduce a major revision within two years of that date, in order to give vendors and users a breathing space to develop, market and implement products. The subsequent specifications 2.1 and 2.2 involve relatively minor changes with which the earlier versions remain upwardly compatible. When Version 3.0 is published it will contain some elements – notably the Manufacturing Messaging Service – which may require some modifications to existing equipment or software if it is to be compatible with new generation products. With the establishment of Version 3.0 it is expected that a longer period of stability will be declared, with upward compatibility maintained with any further developments within that period.

CONFORMANCE TO SPECIFICATION

MAP and TOP represent a breakthrough in collaboration between fiercely competing vendors who have accepted the authority of an independent common standard. It is essential, therefore, that conformance to that standard should be independently verified, both to give assurance to the purchaser of equipment and software and to provide a reference against which vendors can measure their products.

The first organisation to set up testing equipment for

MAP conformance was the Industrial Technology Institute (ITI) in the USA. Its test procedures were used on products to be incorporated in the MAP/TOP demonstration at the Autofact exhibition in Chicago in 1985, and were used again by the conformance testing group for the CIMAP event in Birmingham in 1986. As yet the tests are by no means exhaustive, and the fact that different products conform to the specification does not necessarily mean that they will work together – though the effect of conformance testing is to raise the user's confidence that devices will communicate successfully.

Testing is an obvious necessity, and as more and more companies internationally offer MAP and TOP products the need for test facilities will grow rapidly, far beyond the capacity of ITI to handle. In any case, there will be strong pressure from vendors and others to have local test facilities in each country, or at least in each region, and local test houses will want to set up their own testing methods. So the need is now being appreciated for some international supervision, not only of the conformance of products but of the consistency of test procedures around the world. A number of interests are involved here, ranging from national governments to groups of vendors. Two powerful vendor associations which are concerning themselves with conformance testing and test standards are the Corporation for Open Systems in the USA and the Standards Promotion and Application Group in Europe.

APPENDIX B –

WHAT DOES IT COST TO INSTALL MAP?

Some cost ranges experienced in setting up the demonstration MAP/TOP network for the CIMAP event at Birmingham in December 1986 are shown in Fig. B.1. The figures relate to a very simple network of 2,000 metres with 40 nodes, and requiring only short lengths of cable.

To the cost of the network itself must be added the cost of installing a MAP system on the network. The values in Fig. B.2 are again based on the experience of CIMAP. MAP devices are still relatively expensive by comparison with proprietary network hardware, but prices are falling rapidly as a result of increasing demand, and also because of new options in the latest and forthcoming MAP specifications, such as carrierband MAP and Mini-MAP. A

*	*Network design*	*£1,000-1,500*
*	*Cable and components*	*£5-8 per metre installed*
*	*Drop cables (TOP)*	*£50-70 per drop installed*
*	*Installation and tuning*	*£1,500-1,700*
	Total	*£14,500-22,000*

Fig. B.1. Typical network cost based on the CIMAP network

*	*Head-ends*	*£5,000-7,000*
*	*Broadband nodes*	*£3,000-5,000*
*	*Network managers*	*£15,000-20,000*
*	*Analysers*	*£20,000-30,000*
*	*Token scopes*	*£30,000-32,000*
	A small pilot system might cost	*£60,000-120,000*

Fig. B.2. System costs for MAP, based on the CIMAP network

small MAP pilot system will at present typically demand a minimum of £60,000 as system entry fee, but the cost could easily go as high as £120,000.

The largest unknown in terms of cost – and usually the largest cost element of any implementation – is going to be the cost of systems integration software. Figures will naturally depend on the level of in-house expertise and the complexity of the applications to be linked together. However, using CIMAP as a yardstick this figure could be as much as 60% of total project costs. Fig. B.3 shows a cost breakdown for a typical CIMAP cell.

It is not only in money terms that the system integration is the most demanding. It also requires by far the greatest proportion of the time required for carrying out the implementation, as shown in the diagram of Fig. B.4 based on CIMAP experience. Building the MAP/TOP system is the next largest commitment. Broadband network installation and tuning is the easiest part of the task to achieve.

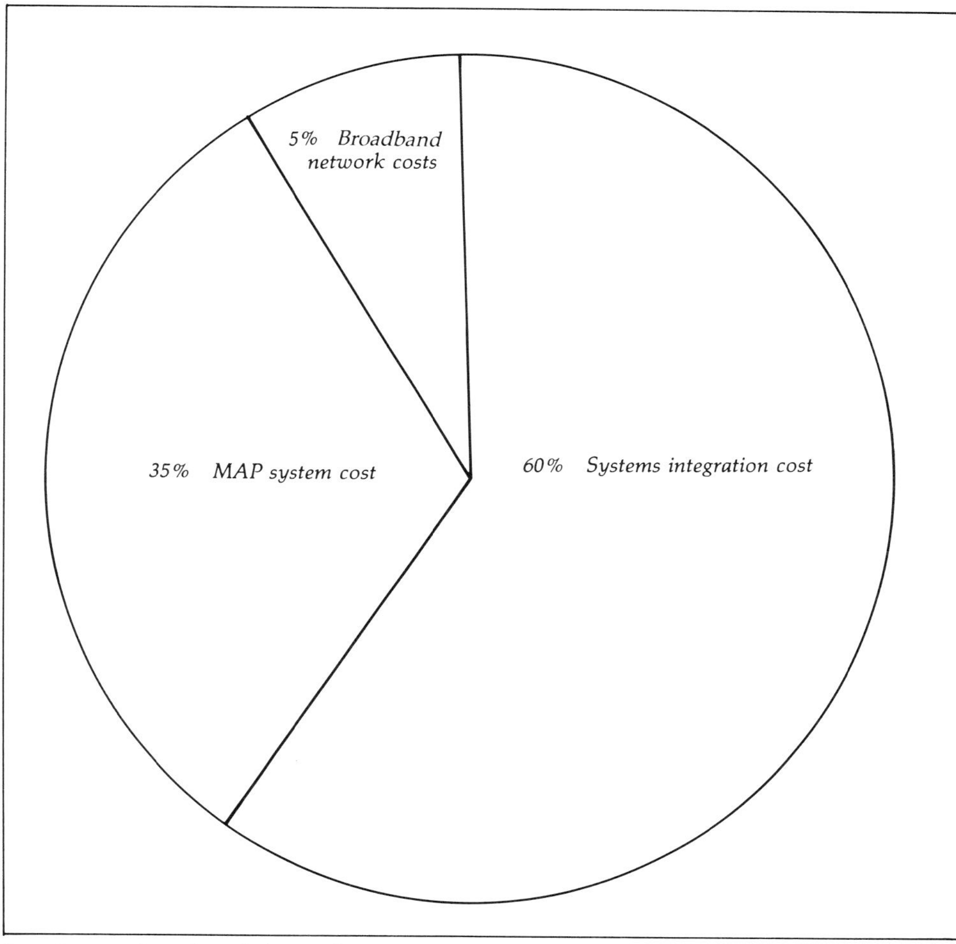

Fig. B.3. Cost breakdown for a typical CIMAP cell

The following figures are estimated from a typical CIMAP demonstration cell:

Network build and tune (entire network): 5-10 man days
MAP device connection and test (entire network): 40-60 man days
Integration software development: 90-120 man days

The costs and timescales for implementation are relatively high, but as more people and organisations

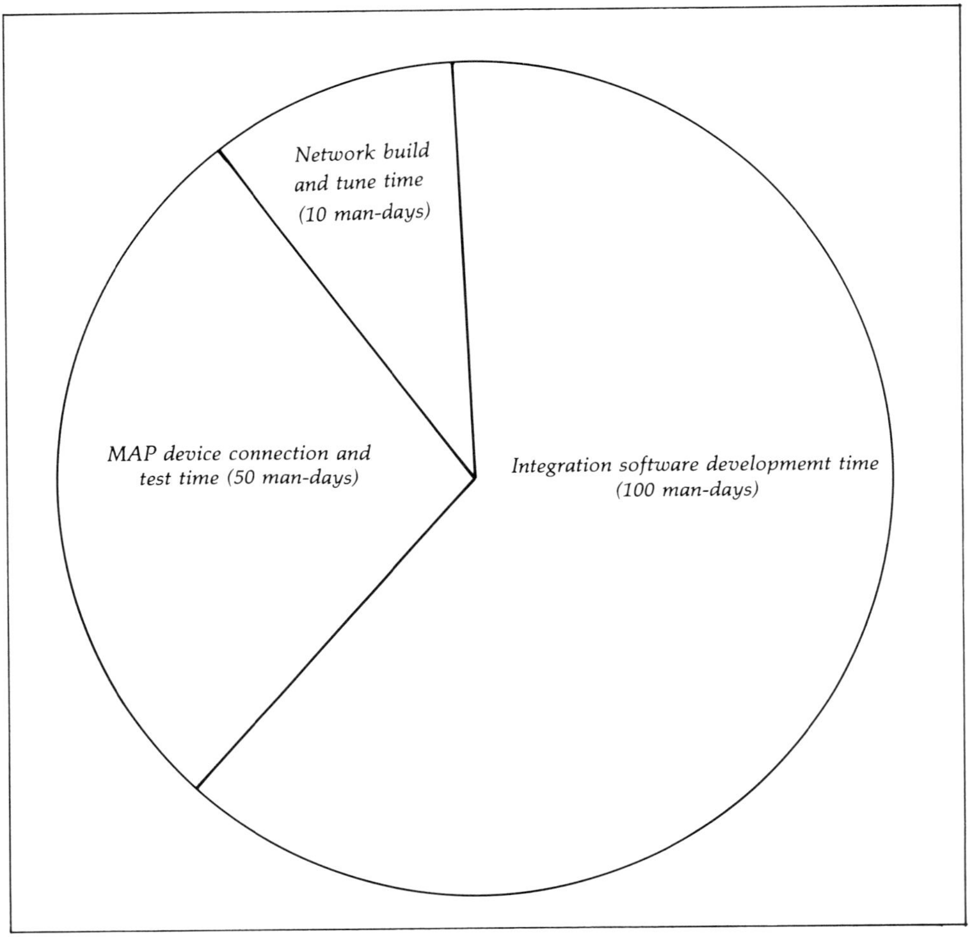

Fig. B.4. Percentage time breakdown for implementing a typical CIMAP cell

acquire the technology and system integration skills these factors will reduce in importance.

BENEFITS

So much for costs. What are the benefits that can be expected in return for this expenditure? The fact must be faced immediately that a first MAP installation project is unlikely in itself to produce significant savings. This is not universally true. Experience has shown that it is possible to achieve a positive return from an initial investment, but it would be misleading to suggest that this is often the case. The strategy for implementing the integrated system must therefore take this probability into account. Once a multivendor MAP/TOP communications network is installed in a factory, successive projects become significantly simpler, less costly in proportion to their size, and more easily justified. A long-term view needs therefore to be taken, and given the fact that MAP and TOP provide a 'utility' this should not be difficult to argue.

What MAP and TOP provide – and this is their biggest benefit – is a common factory utility, similar to three-phase electricity, which will overcome the communications roadblock which companies face in tackling automation projects today. It is therefore best considered as an enabling technology which overcomes many of the communications problems that exist where computers and other equipment from different vendors have to be linked.

Many other specific benefits exist through the use of standardised communications, and some of them are listed in Fig. B.5. In the short term the more realistic benefits can be expected to come from:

- Reduced cabling costs compared with the alternative of multiplying point-to-point links and incompatible networks.
- Freedom to choose the most suitable or least expensive vendor in the purchase of automation equipment.
- Reliability of communications. Communications in a CIM environment are often crucial, as a failure can lead directly to lost production or poor product quality.
- Ease of upgrading. The MAP and TOP specifications pay close attention to the need for a 'migration path' allowing users to maintain compatibility with the latest version.
- Flexibility in routing the network. The bus network design gives virtually unlimited freedom in numbers and locations of connections to the network.

* *Network*
- *reduced cable costs*
- *complementary use (e.g. video, voice)*
- *flexible, upgradable*

* *MAP/TOP protocols*
- *wider choice of vendors*
- *upgradable (reduced obsolescence)*
- *formalised communications*

* *Integration*
- *utility simplifies integration*
- *reduced implementation costs*
- *brings computing to factory floor*
- *greater control of resources (men, machines)*
- *reduced training costs*
- *reduced maintenance costs*

Fig. B.5. Benefits from use of MAP and TOP

- Complementary use of video and voice medium applications. This is a special feature of the broadband system which can more than compensate for its extra cost.

In the longer term there are potentially very substantial returns, some of them not so easily measured, the chief of which are:

- Reduced implementation costs, with the emphasis of investment expenditure moving to application software.
- Shorter implementation timescales because of the reduced need for special-purpose integration software.
- Formalised communications procedures for all integration projects, since they all employ a common utility.
- Increased flexibility in the use of computer-based automation, leading to an increase in the level and amount of automation applied in industry.
- Greater control of production through the increased

level of automation, showing itself in greater flexibility and quality, and reduced costs.

- A reduction in training costs for all people involved with automation projects – again because of increased standardisation.
- Lower maintenance costs, especially of software, because there is not the need to keep specialists in each of the proprietary protocols, but also a reduction in hardware cost.
- Increased activity to make incremental movements in a process rather than being trapped at a proprietary protocol release level.

WILL IT PAY ITS WAY?

This is the key question on which decisions about information strategy will hinge. There have been suggestions recently that MAP and TOP are only for the very large companies which are already experiencing serious problems in integrating their systems. We believe that this is misleading, and that small- and medium-sized companies should address themselves to the opportunities which have been described in this book.

The comment of one senior executive at the CIMAP event in the UK puts the situation succinctly. "If you are a £100 million or more turnover company and have good manufacturing control today, you should be saying to your company 'Why are we not using these technologies'. If you are a £50 million turnover company with good manufacturing control, you should be saying 'Why are we not planning to use these technologies?'. If you are a smaller manufacturing company, you should be saying 'Are we watching these technologies and the impact they can have on the business?"

There will be different needs according to the size of the company. A small company may well find it most appropriate to opt for the new carrierband MAP specification, which is less expensive in that it does not require head-end remodulators or frequency-agile modems

– or he may find a proprietary network or simple RS-232 links adequate for his present needs. Any or all of these can be run simultaneously on a broadband cable, which will also allow migration to a full MAP capability later if that becomes necessary. Any company with more than one 'island of information' must study the potential for raising profitability by linking these islands.